RAPID REVISION NOTES

A LEVEL

PURE MATHEMATICS

BY M J POWELL, B.Sc. (Hons.),
Diploma in Education.
Senior Lecturer in Mathematics,
Caerleon College of Education.

General Editor – Professor A. J. B. Robertson,
Professor of Chemistry, King's College,
London

CELTIC
REVISION AIDS

Celtic Revision Aids
30–32 Gray's Inn Road,
London, WC1X 8JL

First Published 1982

ISBN 0 86305 020 4

Composition by
Filmtype Services Limited,
Scarborough, North Yorkshire.

Printed and bound in Great Britain by
Cox & Wyman Ltd, Reading

GENERAL EDITOR'S FOREWORD

The Rapid Revision 'A' Level Series is designed for students preparing for G.C.E. 'A' Level, Scottish Highers, Intermediate University and similar examinations. They are comprehensive and may be used either on their own for revision or as a complement to set books and text books. The notes are organised to help students remember facts and to pin-point areas of difficulty. Practice questions are included at the end of each section with answers to the numerical problems at the end of the text. Where necessary, clearly worked through examples have been included in the text.

I am sure students will find these books very helpful.

A J B Robertson
M.A., Ph.D., D. Sc., C. Chem., F.R.S.C.
Professor of Chemistry,
King's College London,
University of London
Formerly Fellow of St John's College Cambridge

AUTHOR'S FOREWORD

In choosing the material of this book careful attention has been paid to the various mathematics syllabuses of the different Examining Boards for the General Certificate of Education 'A' level examinations. The topics covered in the book represent the common core of essential mathematics given in These syllabuses. The book is also suitable for science and engineering students.

The content of each topic is summarized carefully and worked examples indicate clearly the mathematical principles and the appropriate techniques associated with the topic.

The practice questions set on a topic are designed to include problems similar to the preceding worked examples and include questions similar to those found on current examination papers. Correct answers are given at the end of the book.

The reader is thus able to improve his or her knowledge and understanding of mathematics and develop the ability to answer successfully 'A' level standard questions.

Contents

1 ALGEBRA

The remainder theorem, harder factors

Let $f(x) \equiv a_0x^n + a_1x^{n-1} + a_2x^{n-2} + \dots a_{n-1}x + a_n$ be a polynomial of degree n, i.e.
where n is a positive integer, and the coefficients $a_0, a_1, a_2, \dots, a_n$ are rational numbers,

$$\text{then } f(x) \equiv (x - b)Q(x) + R \text{ where}$$

$Q(x)$ is a polynomial of degree $n - 1$ and b, R are constants.

Important identities

$$a^3 \pm b^3 \equiv (a \pm b)(a^2 \mp ab + b^2)$$

$$a^3 + b^3 + c^3 - 3abc \equiv (a + b + c)(a^2 + b^2 + c^2 - bc - ca - ab)$$

An expression of several variables is said to be **homogeneous** if all its terms are of the same degree.
An expression in a, b and c is said to be **cyclic** if it is unaltered when we write b for a, c for b and a for c. Similarly for expressions in four or more variables.
An expression of several variables is said to be **symmetrical** in these variables if its value is unaltered by the interchange of any two of the variables.

Examples

1 Factorise (i) $54xa^6 - 16x^4b^9$
(ii) $10x^3 - 51x^2 + 38x + 24$

(i) $54xa^b - 16x^4b^9$

$$= 2x(27a^6 - 8x^3b^9)$$

$$= 2x\{(3a^2)^3 - (2xb^3)^3\}$$

$$= \mathbf{2x(3a^2 - 2xb^3)(9a^4 + 6xa^2b^3 + 4x^2b^6)}$$

(ii) $f(x) \equiv 10x^3 - 51x^2 + 38x + 24$
If $ax + b$ is a factor of $f(x)$, then a is a factor of 10 and b is a factor of 24.
Try factors of the form $x \pm b$ first:

$$f(\pm 1) \neq 0, \quad f(\pm 2) \neq 0, \quad f(\pm 3) \neq 0$$

$$f(4) = 10 \cdot 4^3 - 51 \cdot 4^2 + 38 \cdot 4 + 24$$

$$= 640 - 816 + 152 + 24$$

$$= 0$$

$$\therefore \ x - 4 \text{ is a factor}$$

$$
\begin{array}{r}
10x^2 - 11x - 6 \\
x - 4\overline{)\,10x^3 - 51x^2 + 38x + 24} \\
\underline{10x^3 - 40x^2} \\
- 11x^2 + 38x \\
\underline{- 11x^2 + 44x} \\
- 6x + 24 \\
\underline{- 6x + 24} \\
\underline{\cdot \quad \cdot}
\end{array}
$$

$$f(x) \equiv (x - 4)\,(10x^2 - 11x - 6)$$
$$\equiv \mathbf{(x - 4)\,(2x - 3)\,(5x + 2)}$$

2 Factorise $a(b - c)^3 + b(c - a)^3 + c(a - b)^3$

$$f(a,b,c) \equiv a(b - c)^3 + b(c - a)^3 + c(a - b)^3$$
$$\text{when } b = a, f(a,a,c) = a(a - c)^3 + a(c - a)^3 + c(a - a)^3$$
$$\therefore\ a - b \text{ is a factor}$$

Similarly $(b - c)$ and $(c - a)$ are factors, but $f(a, b, c)$ is homogeneous and of the fourth degree so the other factor must be cyclic, homogeneous and of the first degree
i.e. $k(a + b + c)$ where k is a constant.

$\therefore\ a(b - c)^3 + b(c - a)^3 + c(a - b)^3 \equiv k(a + b + c)(a - b)(c - c)(c - a)$
Comparing coefficients of ab^3 on both sides

$$1 = k$$

$$\text{hence } f(a,b,c) \equiv \mathbf{(a + b + c)\,(a - b)\,(b - c)\,(c - a)}$$

Practice questions 1

1 Factorise (i) $125x^3 + 64y^6$
(ii) $81a^3b^2 - 648b^2c^6$

2 Factorise (i) $x^3(y - z) + y^3(z - x) + z^3(x - y)$
(ii) $a(b^4 - c^4) + b(c^4 - a^4) + c(a^4 - b^4)$

3 Factorise (i) $2x^3 + 3x^2 - 17x - 30$
(ii) $24x^3 - 10x^2 - 7x + 1$

4 Use the remainder theorem to find the remainder when

$$6x^3 - 11x^2 - 19x + 2$$

is divided by $2x + 1$.
Hence, or otherwise, solve the equation

$$6x^3 - 11x^2 - 19x - 6 = 0$$

5 Prove that a repeated linear factor of a polynomial $f(x)$ is also a factor of $f'(x)$
Factorise $x^4 + 2x^3 - 4x^2 - 10x - 5$, given that this expression has a repeated linear factor.
Check your result by multiplying your factors together.

Relations between the coefficients and the roots of an algebraic equation

(i) quadratic equation

$$ax^2 + bx + c = 0$$

then
$$x = \frac{-b \pm \sqrt{b^2 - 4ac}}{2a}$$

If the two values of x are x_1 and x_2 then

$$x_1 + x_2 = -\frac{b}{a} \quad \text{and} \quad x_1 x_2 = \frac{c}{a}$$

(ii) cubic equations

$$ax^3 + bx^2 + cx + d = 0$$

Let the roots be x_1, x_2, x_3 then

$$x_1 + x_2 + x_3 = -\frac{b}{a}$$

$$x_1 x_2 + x_2 x_3 + x_3 x_1 = \frac{c}{a}$$

$$x_1 x_2 x_3 = -\frac{d}{a}$$

(iii) equations of the nth degree

$$a_0 x^n + a_1 x^{n-1} + a_2 x^{n-2} + \ldots + a_{n-1} x + a_n = 0$$

Let the roots be $x_1, x_2, x_3, \ldots, x_n$ then

the sum of the roots, $x_1 + x_2 + \ldots + x_n = \frac{a_1}{a_0}$

the sum of the product of the roots in pairs, $x_1 x_2 + x_2 x_3 + \ldots = \frac{a_2}{a_0}$

the sum of the products of the roots in threes $= -\frac{a_3}{a_0}$

and so on

the product of the roots, $x_1x_2x_3\ldots x_n = (-1)^n\dfrac{a_n}{a_0}$

By means of these relations any symmetrical function of the roots of an equation may be expressed in terms of the coefficients.

The equation whose roots are $\dfrac{1}{x_1}, \dfrac{1}{x_2}, \dfrac{1}{x_3}, \ldots, \dfrac{1}{x_n}$ is

$$a_nx^n + a^{n-1}x^{n-1} + \ldots + a_0 = 0$$

Examples

1 If α, β, γ are the roots of the equation $2x^3 - 5x^2 - 7x + 4 = 0$ find (a) the value of $\alpha^2 + \beta^2 + \gamma^2$
(b) the equation whose roots are α^2, β^2, and γ^2

$$2x^3 - 5x^2 - 7x + 4 = 0$$

$$\alpha + \beta + \gamma = \frac{5}{2} \qquad \text{(i)}$$

$$\alpha\beta + \beta\gamma + \gamma\alpha = -\frac{7}{2} \qquad \text{(ii)}$$

$$\alpha\beta\gamma = -\frac{4}{2} = -2 \qquad \text{(iii)}$$

$$\alpha^2 + \beta^2 + \gamma^2 = (\alpha + \beta + \gamma)^2 - 2(\alpha\beta + \beta\gamma + \gamma\alpha)$$

$$= \left(\frac{5}{2}\right)^2 - 2\left(-\frac{7}{2}\right) \text{ using (i) and (ii)}$$

$$= 6\tfrac{1}{4} + 7 = 13\tfrac{1}{4}$$

$$\alpha^2\beta^2 + \beta^2\gamma^2 + \gamma^2\alpha^2 = (\alpha\beta + \beta\gamma + \gamma\alpha)^2 - 2\alpha\beta\gamma(\alpha + \beta + \gamma)$$

$$= \left(-\frac{7}{2}\right)^2 - 2(-2)\left(\frac{5}{2}\right) \text{ using (i), (ii) and (iii)}$$

$$= \frac{49}{4} + 10$$

$$= 12\tfrac{1}{4} + 10$$

$$= 22\tfrac{1}{4}$$

$$\alpha^2\beta^2\gamma^2 = 4 \text{ (from (iii))}$$

hence the equation whose roots are α^2, β^2, γ^2 is

$$x^3 - 13\tfrac{1}{4}x^2 + 22\tfrac{1}{4}x - 4 = 0$$

i.e. $$4x^3 - 53x^2 + 89x - 16 = 0$$

2 If x_1, x_2, x_3 are the roots of the equation $x^3 + 5x^2 + 8x - 2 = 0$ find (a) $x_1^3 + x_2^3 + x_3^3$ and (b) $x_1^5 + x_2^5 + x_3^5$

$$x^3 + 5x^2 + 8x - 2 = 0$$

$$x_1 + x_2 + x_3 = -5 \quad \text{(i)}$$

$$x_1x_2 + x_2x_3 + x_3x_1 = 8 \quad \text{(ii)}$$

$$x_1x_2x_3 = -2 \quad \text{(iii)}$$

the equation whose roots are $\frac{1}{x_1}, \frac{1}{x_2}, \frac{1}{x_3}$ is

$$-2x^3 + 8x^2 + 5 + 1 = 0$$

$$\therefore \frac{1}{x_1} + \frac{1}{x_2} + \frac{1}{x_3} = -\frac{8}{-2} = 4 \quad \text{(iv)}$$

Since x_1 is a root of the equation $x^3 + 5x^2 + 8x - 2 = 0$

$$x_1^3 + 5x_1^2 + 8x_1 - 2 = 0 \quad \text{(v)}$$

dividing by x_1,

$$x_1^2 + 5x_1 + 8 - \frac{2}{x_1} = 0$$

similarly $$x_2^2 + 5x_2 + 8 - \frac{2}{x_2} = 0$$

and $$x_3^2 + 5x_3 + 8 - \frac{2}{x_3} = 0$$

adding

$$(x_1^2 + x_2^2 + x_3^2) + 5(x_1 + x_2 + x_3) + 24 - 2\left(\frac{1}{x_1} + \frac{1}{x_2} + \frac{1}{x_3}\right) = 0$$

$$(x_1^2 + x_2^2 + x_3^2) + 5(-5) + 24 - 2(4) = 0 \quad \text{using (i) and (iv)}$$

$$x_1^2 + x_2^2 + x_3^2 - 25 + 24 - 8 = 0$$

$$x_1^2 + x_2^2 + x_3^2 = 9 \quad \text{(vi)}$$

rewriting (v)

$$x_1^3 + 5x_1^2 + 8x_1 - 2 = 0$$

similarly $x_2^3 + 5x_2^2 + 8x_2 - 2 = 0$

and $x_3^3 + 5x_3^2 + 8x_3 - 2 = 0$

adding $(x_1^3 + x_2^3 + x_3^3) + 5(x_1^2 + x_2^2 + x_3^2) + 8(x_1 + x_2 + x_3) - 6 = 0$

$(x_1^3 + x_2^3 + x_3^3) + 5(9) + 8(-5) - 6 = 0$ using (i) and (vi)

$\mathbf{x_1^3 + x_2^3 + x_3^3 = 1}$ (vii)

multiplying (v) by x_1 we get

$$x_1^4 + 5x_1^3 + 8x_1^2 - 2x_1 = 0$$

similarly $x_2^4 + 5x_2^3 + 8x_2^2 - 2x_2 = 0$

and $x_3^4 + 5x_3^3 + 8x_3^2 - 2x_3 = 0$

adding $x_1^4 + x_2^4 + x_3^4 + 5(x_1^3 + x_2^3 + x_3^3) + 8(x_1^2 + x_2^2 + x_3^2)$

$- 2(x_1 + x_2 + x_3) = 0$

or using the Σ notation this becomes

$$\Sigma x_1^4 + 5\Sigma x_1^3 + 8\Sigma x_1^2 - 2\Sigma x_1 = 0$$

$\Sigma x_1^4 + 5(1) + 8(9) - 2(-5) = 0$ using (vii) (vi) & (i)

$\Sigma x_1^4 = -87$ (viii)

multiplying (v) by x_1^2 we get

$$x_1^5 + 5x_1^4 + 8x_1^3 - 2x_1^2 = 0$$

similarly $x_2^5 + 5x_2^4 + 8x_2^3 - 2x_2^2 = 0$

and $x_3^5 + 5x_3^4 + 8x_3^3 - 2x_3^2 = 0$

adding $\Sigma x_1^5 + 5\Sigma x_1^4 + 8\Sigma x_1^3 - 2\Sigma x_1^2 = 0$

$\Sigma x_1^5 + 5(-87) + 8(1) - 2(9) = 0$ using (viii), (vii) & (vi)

$$\Sigma x_1^5 = \mathbf{x_1^5 + x_2^5 + x_3^5 = 445}$$

Practice questions 2

1 If α, β, γ are the roots of the equation $2x^3 - 3x + 1 = 0$, find (i) $\Sigma\alpha^2$ (ii) the equation whose roots are α^2, β^2 and γ^2.

2 If x_1, x_2, x_3 are the roots of the equation $x^3 - x^2 + 2x - 3 = 0$, find the values of $x_1^3 + x_2^3 + x_3^3$ and $x_1^4 + x_2^4 + x_3^4$.

3 If α, β, γ, are the roots of the equation $x^3 - x^2 - 4 = 0$, find the equation whose roots are: $\alpha + \beta, \beta + \gamma, \gamma + \alpha$.

4 If the roots of the equation $x^3 - 24x - 84 = 0$ are α, β, γ, find the equation whose roots are $\alpha + 3, \beta + 3, \gamma + 3$. Show that the equation $x^3 - 24x - 84 = 0$ has only one real root and find its value correct to 1 decimal place.

5 If the roots of the equation $x^3 - x^2 + 4x + 3 = 0$ are x_1, x_2, x_3, find the equation whose roots are $2x_1 + x_2 + x_3$, $2x_2 + x_3 + x_1$, $2x_3 + x_1 + x_2$.

Partial fractions

I Consider proper fractions of the form $N(x)/D(x)$ where $N(x)$ and $D(x)$ are polynomials in x and the degree of N is at least one less than the degree of D. Then:

1 To each non-repeated linear factor $(x - a)$ of D there corresponds a partial fraction of the form $A/(x - a)$ where A is a non-zero constant.

2 To each repeated linear factor $(x - a)^2$ of D there corresponds partial fractions of the form $A/(x - a) + B/(x - a)^2$, where A and B are non-zero constants.

3 To each repeated linear factor $(x - a)^r$ of D there corresponds r partial fractions of the form

$$\frac{A_1}{x - a} + \frac{A_2}{(x - a)^2} + \frac{A_3}{(x - a)^3} + \ldots + \frac{A_r}{(x - a)^r},$$

where $A_1, A_2, \ldots, A_r$ are constants.

4 To each quadratic factor $ax^2 + bx + c$ of D which does not factorise into two linear factors there corresponds a partial fraction of the form

$$\frac{Ax - B}{ax^2 + bx + c}$$

where A and B are constants.

5 To each repeated quadratic factor $(ax^2 + bx + c)^r$ of D there corresponds r partial fractions of the form

$$\frac{A_1x + B_1}{ax^2 + bx + c} + \frac{A_2x + B_2}{(ax^2 + bx + c)^2} + \ldots + \frac{A_1x + B_3}{(ax^2 + bx + c)^r}$$

where As and Bs are constants.

II Fractions of the form $N(x)/D(x)$ where the degree of N is equal to or higher than the degree of D. First divide N by D obtaining a polynomial in x, $P(x)$, plus a remainder $R(x)$ where R is of lower degree than D.
Then $N(x)/D(x) = P(x) + R(x)/D(x)$ and the rules of I above can then be applied to $R(x)/D(x)$.

Examples

1 Express in terms of partial fractions

(i) $\dfrac{4x^2 - 11x + 3}{x^3 - 2x^2 - x + 2}$ (ii) $\dfrac{3x^2 - 6x + 13}{(x-3)(x^2 + x - 1)}$

(i)
$$f(x) \equiv x^3 - 2x^2 - x + 2$$
$$f(2) = 2^3 - 2^3 - 2 + 2 = 0$$
$$\therefore\ x - 2 \text{ is a factor}$$
$$f(1) = 1 - 2 - 1 + 2 = 0$$
$$\therefore\ x - 1 \text{ is a factor}$$
$$f(-1) = -1 - 2 + 1 + 2 = 0$$
$$\therefore\ x + 1 \text{ is a factor}$$
$$\therefore\ x^3 - 2x^2 - x + 2 = k(x-1)(x+1)(x-2)$$

comparing coefficients of x^3, $k = 1$

$$\therefore\ x^3 - 2x^2 - x + 2 \equiv (x-1)(x+1)(x-2)$$

$$\frac{4x^2 - 11x + 3}{x^3 - 2x^2 - x + 2} \equiv \frac{A}{x-1} + \frac{B}{x+1} + \frac{C}{x-2}$$

$$4x^2 - 11x + 3 \equiv A(x+1)(x-2) + B(x-1)(x-2) + C(x-1)(x+1)$$

Put $x = 1$, $\quad -4 = A(2)(-1)$
$$A = 2$$
Put $x = -1$, $\quad 18 = B(-2)(-3)$
$$B = 3$$
Put $x = 2$, $\quad -3 = C(1)(3)$
$$C = -1$$

$$\therefore\ \frac{4x^2 - 11x + 3}{x^3 - 2x^2 - x + 2} \equiv \frac{2}{x-1} + \frac{3}{x+1} - \frac{1}{x-2}$$

(ii)
$$\frac{3x^2 - 6x + 13}{(x-3)(x^2 + x - 1)} \equiv \frac{A}{x-3} + \frac{Bx + C}{x^2 + x - 1}$$
$$3x^2 - 6x + 13 \equiv A(x^2 + x - 1) + (Bx + C)(x - 3)$$

Put $x = 3$, $\quad 22 = A11$
$$A = 2$$
$$\therefore\ 3x^2 - 6x + 13 \equiv 2(x^2 + x - 1) + (Bx + C)(x - 3)$$

comparing coefficients of x^2, $\quad 3 = 2 + B$

$$B = 1$$

comparing constant terms $\qquad 13 = -2 - 3C$

$$C = -5$$

$$\therefore \quad \frac{3x^2 - 6x + 13}{(x-3)(x^2+x-1)} = \frac{2}{x-3} + \frac{x-5}{x^2+x-1}$$

2 Show that $x + 2$ is a factor of $x^3 + 8x^2 + 21x + 18$ and hence factorise the expression completely.
Express in terms of partial fractions

$$\frac{3x^2 + 15x + 19}{x^3 + 8x^2 + 21x + 18}$$

$$f(x) \equiv x^3 + 8x^2 = 21x + 18$$

$$f(-2) = -8 + 32 - 42 + 18 = 0$$

$$\therefore \; x + 2 \text{ is a factor}$$

$$\begin{array}{r|l}
 & x^2 + 6x + 9 \\
\cline{2-2}
x + 2 & x^3 + 8x^2 + 21x + 18 \\
 & x^3 + 2x^2 \\
\cline{2-2}
 & \quad 6x^2 + 21x \\
 & \quad 6x^2 + 12x \\
\cline{2-2}
 & \qquad 9x + 18 \\
 & \qquad 9x + 18 \\
\cline{2-2}
\end{array}$$

$$f(x) = (x+2)(x + 6x + 9)$$

$$\equiv (x+2)(x+3)^2$$

$$\frac{3x^2 + 15x + 19}{x^3 + 8x^2 + 21x + 18} \equiv \frac{A}{x+2} + \frac{B}{x+3} + \frac{C}{(x+3)^2}$$

$$3x^2 + 15x + 19 \equiv A(x+3)^2 + B(x+2)(x+3) + C(x+2)$$

Put $x = -2$

$$12 - 30 + 19 = A$$

$$A = 1$$

Put $x = -3$

$$27 - 45 + 19 = -C$$

$$C = -1$$

Comparing coefficients of x^2

$$3 = A + B$$
$$3 = 1 + B$$
$$B = 2$$

$$\therefore \frac{3x^2 + 15x + 19}{x^3 + 8x^2 + 21x + 18} \equiv \frac{1}{x+2} + \frac{2}{x+3} - \frac{1}{(x+3)^2}$$

3 Express in terms of partial fractions

$$\frac{x^4}{(x+1)(x^2+1)}$$

We need to find the quotient only until the remainder is a quadratic.

$$\begin{array}{r|l} & x - 1 \\ \hline x^3 + x^2 + x + 1 & x^4 \\ & x^4 + x^3 \ldots \\ \hline & -x^3 \ldots \\ & -x^3 \ldots \\ \hline & \text{quadratic remainder} \\ \hline \end{array}$$

$$\frac{x^4}{(x+1)(x^2+1)} \equiv x - 1 + \frac{A}{x+1} + \frac{Bx + C}{x^2+1}$$

$$x^4 \equiv (x-1)(x+1)(x^2+1) + A(x^2+1) + (Bx + C)(x+1)$$

Put $x = -1$

$$1 = 2A$$
$$A = \tfrac{1}{2}$$

Compare constant terms

$$0 = -1 + A + C$$
$$0 = -1 + \tfrac{1}{2} + C$$
$$C = \tfrac{1}{2}$$

Put $x = 1$

$$1 = 2A + 2(B + C)$$
$$1 = 1 + 2(B + \tfrac{1}{2})$$
$$B = -\tfrac{1}{2}$$

$$\frac{x^4}{(x+1)(x^2+1)} \equiv x - 1 + \frac{1}{2(x-1)} + \frac{1-x}{2(x^2+1)}$$

Practice questions 3

1 Express in terms of partial fractions

(i) $\dfrac{5x^2+12x+4}{(2x+1)(x^2+3x+1)}$ (ii) $\dfrac{20x^2-34x+8}{3x^3-10x^2+9x-2}$

(iii) $\dfrac{6x^2-23x+30}{(x+2)(x-3)^2}$

2 Express in terms of partial fractions

(i) $\dfrac{2x^3+x^2-20x+11}{x^2+x-12}$

(ii) $\dfrac{x^3+5x^2+3x+4}{(x+2)(x^2+1)}$

3 Express in terms of partial fractions

$$\frac{x^3}{(x-1)(x^2+x+4)}$$

Check your result by reducing your partial fractions to a single fraction.

Permutations

The number of permutations of n objects taken r at a time (written ${}^nP_r$ or nPr) is the number of ways of choosing r objects from the n objects, different orders in which the r objects are chosen counting as different permutations. In choosing the r objects, we have a choice of n objects for the first one chosen, then a choice of $(n-1)$ objects for the second one chosen, and so on. Hence

$$\begin{aligned} {}^nP_r &= n(n-1)(n-2)\ldots(n-r+1) \\ &= n(n-1)(n-2)\ldots(n-r+1) \times \frac{(n-r)(n-r-1)\ldots2\cdot1}{(n-r)(n-r-1)\ldots2\cdot1} \\ &= \frac{n!}{(n-r)!} \end{aligned}$$

The number of permutations of n objects taken n at a time is given by $n(n-1)(n-2)\ldots2\cdot1 = n!$
Putting $r = n$ in the general result gives us $n!/0!$ so $0! = 1$.

The number of permutations of n objects taken n at a time when a of the objects are identical is given by

$$\frac{n!}{a!}$$

Examples

1 Find the numbers of ways of arranging 10 people
 (a) to stand in a single file queue
 (b) to sit at a circular table

 (a) Number of arrangements is $10\cdot 9 \ldots 2\cdot 1 = 10!$
 (b) In this case there is no beginning or end. We have to consider the arrangement of the people relative to each other. So choose one person. Then arrange the 9 others about him. Number of arrangements is therefore 9!

2 How many numbers greater than 300 can be formed from the digits 2, 3, 4, 5?
 Consider the 3 digit numbers which can be formed.
 The hundreds place can be filled by either 3, 4 or 5 i.e. in 3 ways.
 Having filled the hundreds place the tens place can be filled in 3 ways and then the units placed in 2 ways.
 Hence the total number of 3 digit numbers over 300 which can be found is $3 \times 3 \times 2 = 18$.
 Consider the 4 digit numbers which can be formed.
 The thousands place can be filled in 4 ways then the hundreds place in 3 ways and so on giving $4 \times 3 \times 2 \times 1 = 24$ numbers.
 So the total number of numbers over 300 which can be formed is $18 + 24 = 42$.

Combinations

The number of combinations of n objects taken r at a time (written ${}^nC_r$, or nCr) is the number of ways of choosing a set of r objects from the n objects, the order in which the r objects are chosen not being significant i.e. different orders of selection for the same objects count as one combination. So ${}^rP_r$ permutations count as one combination. Hence

$${}^nC_r = \frac{{}^nP_r}{{}^rP_r} = \frac{{}^nP_r}{r!}$$

$${}^nC_r = \frac{n!}{(n-r)!r!}$$

It follows that

$${}^nC_{n-r} = \frac{n!}{r!(n-r)!}$$

$$\therefore {}^nC_r = {}^nC_{n-r}$$

$$^nC_r + {}^nC_{r+1} = \frac{n!}{(n-r)!r!} + \frac{n!}{(n-r-1)!(r+1)}$$

$$= n!\frac{r+1+n-r}{(n-r)!(r+1)!}$$

$$= \frac{(n+1)!}{(n-r)!(r+1)!}$$

$$\therefore\ {}^n\mathrm{C}_r + {}^n\mathrm{C}_{r+1} = {}^{n+1}\mathrm{C}_{r+1}$$

This result is known as **Vandermonde's Theorem**

N.B.

$$^nC_r = \frac{n!}{(n-r)!r!}$$

$$= \frac{n(n-1)(n-2)\dots(n-r+1)}{1.\ 2.\ 3\dots r}$$

Both numerator and denominators contain r factors.

Examples

1 In how many ways can 8 people be paired?
Here we require the number of ways of selecting 2 people from 8 people.
A, B is the same selection as *B, A*

Number of possible pairs $= {}^8C_2$

$$= \frac{8\cdot7}{1\cdot2}$$

$$= \mathbf{28}$$

2 Evaluate $^{10}C_7$

$$^{10}C_7 = {}^{10}C_3$$

$$= \frac{10.9.8}{1.2.3} = \mathbf{120}$$

3 What is the total number of selections which may be made from n objects if any number may be taken at a time?
Each object may be taken or left.
Hence the total number of selections is 2.2.2.... to n factors i.e. 2^n.
But this includes the case when all are left, so the total number of selections is

$$\mathbf{2^n - 1}$$

also
The number of ways of choosing 1 object is $^nC_1$
The number of ways of choosing 2 objects is $^nC_2$
and so on.

$$^n\mathrm{C}_1 + {}^n\mathrm{C}_2 + {}^n\mathrm{C}_3 \dots + {}^n\mathrm{C}_n = 2^n - 1$$

Practice questions 4

1 How many 4 digit numbers can be formed from the digits 1, 2, 3, 4, 5, 6, each digit to be used once only in forming any number?

2 Two soccer elevens are to be made up from 22 players, only two of whom are goalkeepers. How many arrangements are possible? Give your answer in terms of factorials.

3 In how many ways can three cards be selected from a pack of 52 cards, if at least one of them is a king?

4 Each of 10 children seated around a round table at a party is to be given a red, green or yellow hat. No child is to have the same coloured hat as either of his or her neighbours. In how many different ways can this be done?

Sequences and finite series

A function whose domain is the set of positive integers 1, 2, 3, 4,..., and whose range is a subset of the real numbers is called a **sequence**. Examples of sequences are:

1, 4, 9, 16, 25,...

2, 4, 6, 8, 10,...

1, 1, 2, 3, 5, 8,...

Each element of a sequence is called a **term**.
The general term of a sequence is often denoted by u_r, (the rth term) or u_n, (the nth term). The general sequence is denoted by:

$$u_1, \quad u_2, \quad u_3, \quad \ldots, \quad u_r, \ldots$$

A relation connecting two or more general terms of a sequence is called a recurrence relation, e.g.

$$u_r = u_{r-1} - u_{r-2}, \quad r > 2$$

Given also that the first and second terms of the sequence are 10 and 8 respectively we can determine the other terms of the series in turn,

$$10, 8, -2, -10, -8, 2, 10, 8 \ldots$$

If the terms of a sequence are connected by + signs we form a **series**. The series formed from the sequence $u_1, u_2, u_3, \ldots, u_n$ is

$$u_1 + u_2 + u_3 + \ldots + u_n$$

and is written

$$\sum_{r=1}^{n} u_r$$

The **Arithmetic series,** also called the **Arithmetic Progression,** (A.P.) is of the form

$$a + (a + d) + (a + 2d) + (a + 3d) + \ldots$$

The nth term is $[a + (n - 1)d]$. d is called the **common difference.**
The sum of the first n terms is

$$\sum_{r=1}^{n} [a + (n - 1)d] = \frac{n}{2}[2a + (n - 1)d]$$

The **Geometric series,** also called the **Geometric Progression** (G.P.), is of the form

$$a + aR + aR^2 + aR^3 + \ldots$$

The nth term is aR^{n-1}. R is called the **common ratio.**
The sum of the first n term is

$$\sum_{r=1}^{n} aR^{r-1} = \frac{a(1 - R^n)}{1 - R} \quad \text{(used when } R < 1\text{)}$$

$$= \frac{a(R^n - 1)}{R - 1} \quad \text{(used when } R > 1\text{)}$$

Sums of the Powers of the first n *natural numbers*

$$1 + 2 + 3 + \ldots + n = \sum_{r=1}^{n} r = \tfrac{1}{2}n(n + 1)$$

$$1^2 + 2^2 + 3^2 + \ldots + n^2 = \sum_{r=1}^{n} r^2 = \tfrac{1}{6}n(n + 1)(2n + 1)$$

$$1^3 + 2^2 + 3^3 + \ldots + n^3 = \sum_{r=1}^{n} r^3 = \{\tfrac{1}{2}n(n + 1)\}^2$$

$$\sum_{r=1}^{n} r^3 = \left\{\sum_{r=1}^{n} r\right\}^2$$

Summation of finite series
If it is possible to express the rth term, u_r, as a difference in the form $f(r + 1) - f(r)$ then the sum of the series to n terms can be found.

Example
Find the sum to n terms of the series

$$1.2 + 2.3 + \ldots + n(n + 1)$$

$$u_r = r(r + 1)$$

Consider the expression

$$r(r + 1)(r + 2) - (r - 1)r(r + 1)$$

$$= r(r+1)[(r+2)-(r-1)]$$
$$= 3r(r+1)$$
$$\therefore\ r(r+1) = \tfrac{1}{3}\{r(r+1)(r+2) - (r-1) + (r+1)\}$$
$$[u_r = f(r+1) - f(r)]$$

Let r take successively the values 1, 2, 3..., n in this identity we obtain

$$r = 1,\quad 1.2 = \tfrac{1}{3}\{1.2.3 - 0\}$$
$$r = 2,\quad 2.3 = \tfrac{1}{3}\{2.3.4 - 1.2.3\}$$
$$r = 3,\quad 3.4 = \tfrac{1}{3}\{3.4.5 - 2.3.4\}$$

....

$$r = n-1,\quad (n-1)n = \tfrac{1}{3}\{(n-1)n(n+1) - (n-2)(n-1)n\}$$
$$r = n,\quad n(n+1) = \tfrac{1}{3}\{n(n+1)(n+2) - (n-1)n(n+1)\}$$

Adding we seen that terms cancel out in pairs on the right hand side leaving

$$1.2 + 2.3 + \ldots + n(n+1) = \tfrac{1}{3}\{n(n+1)(n+2) - 0\}$$
$$= \tfrac{1}{3}\mathbf{n(n+1)(n+2)}$$

Practice questions 5

1 Find the sum of the first n terms of the series

$$1^2.4 + 2^2.5 + 3^2.6 + \ldots$$

2 Find the sum to n terms of the series

$$1.3 + 3.5 + 5.7 + \ldots$$

3 Find the sum of the series

$$1.2.3 + 2.3.4 + 3.4.5 + \ldots + n(n+1)(n+2)$$

4 Find the sum of the first n terms of the series whose rth term is

$$r(r+1)(r+2)(r+3)$$

5 Find the sum to n terms of the series

$$\frac{1}{1.3} + \frac{1}{3.5} + \frac{1}{5.7}$$

6 Find the sum of the first n terms of the series

$$1.2 + 3.2^2 + 5.2^3 + 7.2^4 + \ldots$$

The binomial theorem

When n is a positive integer

$$(a + x)^n = a^n + na^{n-1}x + \frac{n(n-1)}{1.2}a^{n-2}x^2 + \frac{n(n-1)(n-2)}{1.2.3}$$
$$\times\, a^{n-3}x^3 + \ldots + x^n$$
$$= a^n + {}^nC_1a^{n-1}x + {}^nC_2a^{n-2}x^2 + {}^nC_3a^{n-3}x^3 + \ldots + x^n$$

There are $(n + 1)$ terms
The degree of each term is n
The $(r + 1)$ term is ${}^nC_ra^{n-r}x^r$. This is also known as the general term.
Since ${}^nC_{r-1} + {}^nC_r = {}^{n+1}C_r$ (Vandermonde's theorem) the coefficients in the expansion of $(a + x)^{n+1}$ can be obtained from the coefficients of $(a + x)^n$.
The pattern of the coefficients in the expansion of $(a + x)^n$ for $n = 1, 2, \ldots,$ is shown in the following triangle known as Pascal's triangle.

n					coefficients						
1					1		1				
2				1		2		1			
3			1		3		3		1		
4		1		4		6		4		1	
5	1		5		10		10		5		1
						etc.					

Suppose we require the expansion of $(a + x)^4$
From the table when $n = 4$ we get 1 4 6 4 1, hence $(a + x)^4 = a^4 + 4a^3x + 6a^2x + 4a^1x^3 + x^4$
To obtain the row for $n = 6$, start and end the row with a 1.
Terms in between are obtained by adding together the two terms on either side of it in the row above.
Thus when $n = 6$, the row is

1 6 15 20 15 6 1

N.B. the symmetry of each row which arises because

$${}^nC_r = {}^nC_{n-r}$$

Examples
Write down the general term in the expansion of

$$\left(\frac{1}{x} + 3x\right)^{10}$$

Hence determine the constant term in this expansion
The $(r + 1)$th term in the expansion of $(1/x + 3x)^{10}$ is

$${}^{10}C_r\left(\frac{1}{x}\right)^{10-r}(3x)^r$$
$$= {}^{10}C_r \,.\, 3^x \,.\, x^{2r-10}$$

The constant term occurs when $2r - 10 = 0$ i.e. when $r = 5$ the constant term is ${}^{10}C_5 \,.\, 3^5$.

$$= \frac{10.9.8.7.6.3^5}{1.2.3.4.5}$$

$$= 61236$$

Show that $(1 - x)^5 (1 + x)^3 (1 + 2x) \simeq 1 + ax^2$ when x is small, such that x^3 and higher powers of x may be neglected, and evaluate a.

$$(1 - x)^5 (1 + x)^3 (1 + 2x)$$
$$= (1 - x^2)^3 (1 - x)^2 (1 + 2x)$$
$$= (1 - 3x^2 + \text{higher powers of } x)(1 - 2x + x^2)(1 + 2x)$$
$$= (1 - 3x^2 + \text{higher powers of } x)\,(1 - 3x^2 + 2x^3)$$
$$= 1 - 6x^2 + \text{higher powers of } x$$
$$\simeq 1 - 6x^2$$

i.e. $\quad a = -6$

Practice questions 6

1 Write down the term independent of x in the expansion of

$$\left(x^2 + \frac{2}{3x^3}\right)^{10},$$

giving the answer in prime factor form.

2 Find the value of b if the expansion of $(1 - 2x)^4 (1 + bx)^5$ contains no term in x. In this case determine the coefficients of x^2 and x^3.

3 Find the first four terms and the last term of the expansion of $[1 + (n + 1)x] (1 + x)^{n-1}$ as an ascending power series in x.

4 Find, correct to four significant figures, the value of (a) $(1{\cdot}001)^6$, (b) $(9{\cdot}98)^7$ by means of the Binomial Theorem.

5 Find, without table or calculators, the exact value of

$$(2 - \sqrt{3})^5 + (2 + \sqrt{3})^5.$$

The Binomial Series for any rational index

For all rational values of n, (positive or negative)

$$(1 + x)^n = 1 + nx + \frac{n(n-1)}{2!}x^2 + \frac{n(n-1)(n-2)}{3!}x^3 + \ldots \text{when } |x| < 1$$

i.e. the sum of the infinite series on the right hand side is equal to $(1 + x)^n$ for all rational values of n when $|x| < 1$.

Examples

1 Write down the first 4 terms in the expansions in ascending powers of x of

(a) $(1 + x)^{-2}$ when $|x| < 1$
(b) $(2 - 3x)^{+1/2}$ when $|x| < \frac{2}{3}$

(a)

$$(1+x)^{-2} = 1 + (-2)x + \frac{(-2)(-3)}{1.2}x^2 + \frac{(-2)(-3)(-4)}{1.2.3}x^3 + \ldots$$

$$= 1 - 2X + 3X^2 - 4X^3 + \ldots$$

(b) $(2 - 3x)^{1/2}$

$$= 2^{1/2}(1 - \tfrac{3}{2}x)^{1/2}$$

$$= 2^{1/2}\left[1 + \tfrac{1}{2}(-\tfrac{3}{2}x) + \frac{\frac{1}{2}(-\frac{1}{2})}{1.2}(-\tfrac{3}{2}x)^2 + \frac{(\frac{1}{2})(-\frac{1}{2})(-\frac{3}{2})}{1.2.3}(-\tfrac{3}{2}x)^3 + \ldots\right]$$

$$= 2^{1/2}[1 - \tfrac{3}{4}x - \tfrac{9}{32}x^2 - \tfrac{27}{128}x^3 + \ldots]$$

$$= +\sqrt{2} - \frac{3\sqrt{2}}{4}x - \frac{9\sqrt{2}}{32}x^2 - \frac{27\sqrt{2}}{128}x^3 + \ldots$$

2 Expand

$$\frac{5 - 6x - 8x^2}{(1 + 2x)^2(1 - x)}$$

in a series of ascending powers of x as far as the term in x^3. State for what range of values of x the expansion is valid.

$$\frac{5 - 6x - 8x^2}{(1 + 2x)^2(1 - x)} \equiv \frac{A}{1 + 2x} + \frac{B}{(1 + 2x)^2} + \frac{C}{1 - x}$$

where A, B and C are constants to be determined

$$5 - 6x - 8x^2 \equiv A(1 + 2x)(1 - x) + B(1 - x) + C(1 + 2x)^2$$

Put $x = 1$, $\quad -9 = 9C$

$C = -1$

Put $x = -\frac{1}{2}$, $\quad 6 = \frac{3}{2}B$

$B = 4$

Put $x = 0$, $\quad 5 = A + B + C$

$A = 5 - 4 + 1$

$= 2$

Thus

$$\frac{5 - 6x - 8x^2}{(1 + 2x)^2(1 - x)} \equiv \frac{2}{1 + 2x}\;\frac{4}{(1 + 2x)^2} - \frac{1}{1 - x}$$

$$\equiv 2(1 + 2x)^{-1} + 4(1 + 2x)^{-2} - 1(1 - x)^{-1}$$

when $|x| < \frac{1}{2}$, $\quad (1 + 2x)^{-1} = 1 + (-1)(2x) + \dfrac{(-1)(-2)}{1.2}(2x)^2$

$$+\frac{(-1)(-2)(-3)}{1.2.3}(2x)^3+\ldots$$

$$=1-2x+4x^2-8x^3+\ldots$$

when $|x|<\frac{1}{2}$, $\quad (1+2x)^{-2}=1+(-2)(2x)+\frac{(-2)(-3)}{1.2}(2x)^2$

$$+\frac{(-2)(-3)(-4)}{1.2.3}(2x)^3+\ldots$$

$$=1-4x+12x^2-32x^3+\ldots$$

when $|x|<1$, $\quad (1-x)^{-1}=1+(-1)(-x)+\frac{(-1)(-2)}{1.2}(-x)^2$

$$+\frac{(-1)(-2)(-3)}{1.2.3}(-x)+\ldots$$

$$=1+x+x^2+x^3+\ldots$$

Hence when $|x|<\frac{1}{2}$,

$$\frac{5-6x-8x^2}{(1+2x)^2(1-x)}\equiv 2(1-2x+4x^2-8x^3+\ldots)$$

$$+4(1-4x+12x^2-32x^3+\ldots)$$

$$-(1+x+x^2+x^3+\ldots)$$

$$\equiv \mathbf{5-21x+55x^2-145x^3+\ldots}$$

Practice questions 7

1 Write down the first four terms in the expansions, in ascending powers of x, of the following

(a) $(2-5x)^{-1}$ (b) $(1+3x)^{3/2}$ (c) $(1-x)^{-2/3}$

2 Expand

$$\frac{2-5x}{(1-3x)(1-2x)^2}$$

in a series of ascending powers of x as far as the term in x^3. State for what range of values of x the expansions is valid.

2 CALCULUS

Differentiation

The differential coefficient, the derived function, the derivative

If y is a single valued continuous function of x, written $y = f(x)$, then an increment δx in the independent variable x produces a corresponding increment δy in the dependent variable y.

$$y + \delta y = f(x + \delta x)$$

$$y = f(x)$$

subtracting $\quad \delta y = f(x + \delta x) - f(x)$

dividing by δx, $\quad \delta x \neq 0$

$$\frac{\delta y}{\delta x} = \frac{f(x + \delta x) - f(x)}{\delta x}$$

The limit as $\delta x \to 0$ of

$$\frac{f(x + \delta x) - f(x)}{\delta x}$$

is called the differential coefficient of $f(x)$ with respect to x and is denoted by any one of the following:

$$f'(x), \quad \frac{d}{dx}f(x), \quad \frac{dy}{dx}, \quad Df, \quad y_1$$

The differential coefficient is also called the derived function and also the derivative.

Thus $\displaystyle \frac{dy}{dx} = \lim_{\delta x \to 0} \frac{\delta y}{\delta x} = \lim_{\delta x \to 0} \frac{f(x + \delta x) - f(x)}{\delta x} = f'(x)$

using two of the notations for the derivative.

Example
Obtain from first principles the differential coefficient of x^5

$$\text{Let } y = x^5$$

An increment δx in x gives a corresponding increment δy in y

$$y + \delta y = (x + \delta x)^5$$

$$\delta y = (x + \delta x)^5 - x^5$$

dividing by δx

$$\frac{\delta y}{\delta x} = \frac{(x + \delta x)^5 - x^5}{\delta x}, \quad \delta x \neq 0$$

$$= \frac{[x^5 + 5x^4\delta x + 10x^3(\delta x)^2 + \ldots + (\delta x)^5 - x^5]}{\delta x}$$

[expanding $(x + \delta x)^5$ using the binomial theorem]

$= 5x^4 + 10x^3\delta x +$ terms containing $(\delta x)^2$ or higher powers of δx

$$\therefore \lim_{\delta x \to 0} \frac{\delta y}{\delta x} = 5x^4$$

$$\therefore \quad \frac{dy}{dx} = 5x^4$$

Practice questions 1

1 Obtain from first principles the differential coefficient of

(i) x^6 (ii) $6x^4$ (iii) $(3x + 2)^5$

2 u and v are functions of x, δu, δv, the changes in u and v corresponding to a change δx in x. Find an expression for the change in the product uv and hence prove the 'product rule' for differentiation of a product of two functions of x.

Rules for differentiation

Let u and v be functions of x then

1 $\dfrac{d}{dx}(u \pm v) = \dfrac{du}{dx} \pm \dfrac{dv}{dx}$

2 $\dfrac{d}{dx}(au) = a\dfrac{du}{dx}$, where a is a constant

3 The product rule, $\dfrac{d}{dx}(uv) = \dfrac{du}{dx}v + u\dfrac{dv}{dx}$

4 The quotients rule, $\dfrac{d}{dx}\left(\dfrac{u}{v}\right) = \dfrac{v\dfrac{du}{dx} - u\dfrac{dv}{dx}}{v^2}$

5 The function of a function rule, if y is a function of w, where w is a function of x,

$$\frac{dy}{dx} = \frac{dy}{dw} \cdot \frac{dw}{dx}$$

6 $\dfrac{dx}{dy} = \dfrac{1}{\dfrac{dy}{dx}}, \quad \dfrac{dy}{dx} \neq 0$

The product rule can be extended to the product of three or more functions $u, v, w \ldots$ of x

$$\frac{d}{dx}(uvw\ldots) = \frac{du}{dx}vw\ldots + u\frac{dv}{dx}w\ldots + uv\frac{dw}{dx}\ldots \text{etc.}$$

The function of a function rule can be extended thus: if y is a function of w, w a function of z, z a function of x then

$$\frac{dy}{dx} = \frac{dy}{dw}\,\frac{dw}{dz}\,\frac{dz}{dx}$$

Similarly for more complicated 'function of function' situations.

Standard algebraic differential coefficients

y	$\dfrac{dy}{dx}$
a, a constant	0
x^n	nx^{n-1}, n a rational number
$[f(x)]^n$	$n[f(x)]^{n-1}f'(x)$

Examples

1 Differentiate with respect to x,

(i) $3x^4$ (ii) $x^{5/3}$ (iii) $(7x+5)^6$ (iv) $(x^3+3)^{-2}$

(v) $7(2-3x^2)^{-4/5}$

(i) $\dfrac{d}{dx}(3x^4) = 3\dfrac{d}{dx}(x^4) = 3 \cdot 4x^3 = 12x^3$

(ii) $\dfrac{d}{dx}(x^{5/3}) = \frac{5}{3}x^{2/3}$

(iii)
$$\frac{d}{dx}(7x+5)^6$$
$$= \frac{d}{dx}(u^6) \text{ where } u = 7x+5$$
$$= \frac{d}{du}(u^6)\frac{du}{dx} \text{ and } \frac{du}{dx} = 7$$
$$= 6u^5 7$$
$$= 42u^5$$
$$= 42(7x+5)^5$$

This process can be written down directly thus

$$\frac{d}{dx}(7x+5)^6$$
$$= 6(7x+5)^5 . 7$$
$$= 42(7x+5)^5$$

(iv)
$$\frac{d}{dx}(x^3+3)^{-2}$$
$$= \frac{d}{dx}(u^{-2}) \quad \text{where} \quad u = x^3+3$$

$$= \frac{d}{du}(u^{-2})\frac{du}{dx} \quad \text{and} \quad \frac{du}{dx} = 3x^2$$

$$= (-2)u^{-3}\,3x^2$$

$$= -6(x^3+3)^{-3}x^2$$

$$= \mathbf{-6x^2(x^3+3)^{-3}}$$

Or writing down the result directly

$$\frac{d}{dx}(x^3+3)^{-2}$$

$$= -2(x^3+3)^{-3}.\,3x^2$$

$$= \mathbf{-6x^2(x^3+3)^{-3}}$$

(v)
$$\frac{d}{dx}7(2-3x^2)^{-4/5}$$

$$= 7(-\tfrac{4}{5})(2-3x^2)^{-9/5}(-6x)$$

$$= \mathbf{\frac{168}{5}x(2-3x^2)^{-9/5}}$$

2 Find the differential coefficients w.r.t. to x of

(i) $(2x^2-7)(3x^3-2)$ (ii) $\dfrac{x^2-1}{\sqrt{x^2+2x-1}}$

(i) $y = (2x^2-7)(3x^3-2)$

$$\frac{dy}{dx} = 4x(3x^3-2) + (2x^2-7)9x^2 \qquad \text{using the product rule}$$

$$= 12x^4 - 8x + 18x^4 - 63x^2$$

$$= \mathbf{30x^4 - 63x^2 - 8x}$$

(ii) $y = \dfrac{x^2-1}{\sqrt{x^2+2x-1}}$

$$\frac{dy}{dx} = \frac{\sqrt{x^2+2x-1}\,2x - (x^2-1)\tfrac{1}{2}(x^2+2x-1)^{-1/2}.(2x+2)}{x^2+2x-1}$$

using the quotient rule

$$= \frac{(x^2+2x-1)2x - (x^2-1)(x+1)}{(x^2+2x-1)^{3/2}}$$

$$= \frac{2x^3+4x^2-2x-x^3+x-x^2+1}{(x^2+2x-1)^{3/2}}$$

$$= \frac{2x^3 + 4x^2 - 2x - x^3 + x - x^2 + 1}{(x^2 + 2x - 1)^{3/2}}$$

$$= \frac{\mathbf{x^3 + 3x^2 - x + 1}}{\mathbf{(x^2 + 2x - 1)^{3/2}}}$$

N.B. In both parts the function of a function rule was used directly.

Practice questions 2

1 Differentiate with respect to x

(i) $8x^{-3}$ (ii) $7x^{2/7} + 4x^{3/2}$ (iii) $6(2x^2 + x + 5)^5$

(iv) $(x^2 + 3)(2 - x^3)$ (v) $\dfrac{2x - 1}{x^2 + 2}$

2 Find the differential coefficients with respect to x of

(i) $(3x^2 + 5)(x^3 - 4)(x - 5)^2$

(ii) $\dfrac{2x^2 - 3}{(x^2 + 1)^{1/2}}$

3 Find the derivatives with respect to x of

(i) $(3x - 2)(x + 7)(4 - 5x)$ (ii) $(x^2 + 3x - 2)^{-2/3}$

(iii) $\dfrac{2x^2 + 4x - 7}{(4x - 1)^2}$

Standard trigonometrical differential coefficients

x is in radians

y	$\dfrac{dy}{dx}$
$\sin x$	$\cos x$
$\cos x$	$-\sin x$
$\tan x$	$\sec^2 x$
$\cot x$	$-\text{cosec}^2 x$
$\sec x$	$\sec x \tan x$
$\text{cosec}\, x$	$-\text{cosec}\, x \cot x$

Examples

1 Differentiate with respect to x

(i) $\sin 7x$ (ii) $\cos 3x^2$ (iii) $\sin^4 x$

(iv) $\tan(5x - 2)$ (v) $\sec 9x^2$

(i) $\frac{d}{dx}\sin 7x = 7.\cos 7x$

(ii) $\frac{d}{dx}\cos 3x^2 = -\sin 3x^2 . 6x = -6x\sin 3x^2$

(iii) $\frac{d}{dx}\sin^4 x = 4\sin^3 x.\cos x$

(iv) $\tan(5x-2) = \sec^2(5x-2).5 = 5\sec^2(5x-2)$

(v) $\sec 9x^2 = \sec 9x^2\tan 9x^2 . 18x = 18x\sec 9x^2\tan 9x^2$

2 Differentiate w.r. to x

(i) $\sin 3x\cos x$ (ii) $\tan 2x\sin^2 x$ (iii) $\frac{\cos 2x}{\sin 3x}$

(i) $\frac{d}{dx}\sin 3x\cos x = \cos 3x.3.\cos x + \sin 3x.(-\sin x)$

$= 3\cos 3x\cos x - \sin 3x\sin x$

(ii) $\frac{d}{dx}\tan 2x\sin^2 x = \sec^2 2x.2.\sin^2 x + \tan 2x.2\sin x.\cos x$

$= 2\sec^2 2x\sin^2 x + \tan 2x\sin 2x$

(iii) $\frac{d}{dx}\frac{\cos 2x}{\sin 3x} = \frac{\sin 3x.(-\sin 2x.2) - \cos 2x.(\cos 3x.3)}{\sin^2 3x}$

$= \frac{-2\sin 3x\sin 2x - 3\cos 3x\cos 2x}{\sin^2 3x}$

Practice questions 3

1 Find the differential coefficients w.r. to x of

(i) $\sin 3x$ (ii) $\cos(2x^4)$ (iii) $\tan^4 x$

(iv) $\sec(3x-2)$ (v) $\operatorname{cosec} 7x^3$

2 Differentiate w.r. to x

(i) $\sin 5x\cos 2x$ (ii) $\cot 2x\sin^3 x$

(iii) $\frac{\sin 7x}{\cos 4x}$ (iv) $\sin^4(3x^2-2)$

Standard inverse trigonometrical differential coefficients

y		$\frac{dy}{dx}$
arc sin x,	$\sin^{-1} x$	$\frac{1}{\sqrt{(1-x^2)}}, \quad -\frac{\pi}{2} < y < \frac{\pi}{2}$
arc cos x,	$\cos^{-1} x$	$\frac{-1}{\sqrt{(1-x^2)}}, \quad 0 < y < \pi$
arc tan x,	$\tan^{-1} x$	$\frac{1}{1+x^2}, \quad -\frac{\pi}{2} < y < \frac{\pi}{2}$

Example
Differentiate w.r. to x

(i) $\sin^{-1} 3x$ (ii) $\cos^{-1} 4x^2$ (iii) $\tan^{-1}(x^2+3)$

(i) $\frac{d}{dx}\sin^{-1} 3x = \frac{1}{\sqrt{(1-(3x)^2)}}.3 = \frac{3}{\sqrt{(1-9x^2)}}$

(ii) $\frac{d}{dx}\cos^{-1} 4x^2 = \frac{-1}{\sqrt{(1-16x^4)}}.8x = \frac{-8x}{\sqrt{(1-16x^4)}}$

(iii) $\frac{d}{dx}\tan^{-1}(x^2+3) = \frac{1}{1+(x^2+3)^2}.2x = \frac{2x}{1+(x^2+3)^2}$

$$= \frac{2x}{x^4+6x^2+10}$$

Practice questions 4

1 Differentiate w.r. to x

(i) $\sin^{-1} 5x$ (ii) $\cos^{-1}(3x-1)$ (iii) $\tan^{-1} 8x^3$

2 Find the differential coefficients w.r. to x of

(i) $\sin^{-1}(8x-1)$ (ii) $\cos^{-1}\frac{3x}{2}$ (iii) $\tan^{-1}\frac{5x^2}{3}$

Standard logarithmic and exponential differential coefficients
Logarithms to base e, unless shown otherwise.

y	$\frac{dy}{dx}$
$\log x$	$\frac{1}{x}$
$\log_a x$	$\frac{1}{x}\log_a e$
e^{ax}	ae^{ax}

y	$\dfrac{dy}{dx}$
a^x	$a^x \log a$
$\log f(x)$	$f\left(\dfrac{1}{x}\right).f'(x)$
$e^{f(x)}$	$e^{f(x)}.f'(x)$

$\log_e A$ is also written $\ln A$.

Examples

1 Find the derivatives w.r. to x of

(i) e^{3x} (ii) $\log(7x+5)$

(iii) $e^{2x} + e^{x^2}$ (iv) $\log x^5$

(i) $\dfrac{d}{dx} e^{3x} = 3e^{3x}$

(ii) $\dfrac{d}{dx} \log(7x+5) = \dfrac{7}{7x+5}$

(iii) $\dfrac{d}{dx}(e^{2x} + e^{x2}) = 2e^{2x} + 2xe^{x^2}$

(iv) $\dfrac{d}{dx} \log x^5 = \dfrac{d}{dx}(5 \log x) = \dfrac{5}{x}$

2 Differentiate w.r. to x

(i) $\ln(3x+1)(7x+2)(x^2+4)$ (ii) $\ln\dfrac{8x-3}{7-x}$

(i) $y = \ln(3x+1)(7x+2)(x^2+4)$

$= \ln(3x+1) + \ln(7x+2) + \ln(x^2+4)$

$$\frac{dy}{dx} = \frac{3}{3x+1} + \frac{7}{7x+2} + \frac{2x}{x^2+4}$$

(ii) $y = \ln\dfrac{8x-3}{7-x}$

$= \ln(8x-3) - \ln(7-x)$

$$\frac{dy}{dx} = \frac{8}{8x-3} - \frac{(-1)}{7-x}$$

$$= \frac{8}{8x-3} + \frac{1}{7-x}$$

$$= \frac{56 - 8x + 8x - 3}{(8x - 3)(7 - x)}$$

$$= \frac{53}{(8x - 3)(7 - x)}$$

Practice questions 5

1 Differentiate w.r.t. to x

(i) e^{7x} (ii) $e^{(5x^2+2)}$ (iii) $\ln(8x^2 - 3x + 1)$

(iv) $\ln(7x - 3)(4x - 3)$

2 Differentiate w.r.t. to x

(i) $\ln(7x - 5)(x - 2)(5 - 4x)$

(ii) $\ln \dfrac{3x - 4}{(x - 1)(2x - 3)}$

(iii) $e^{\sin 3x}$

Standard hyperbolic and inverse hyperbolic differential coefficients

y	$\dfrac{dy}{dx}$
$\sinh x$	$\cosh x$
$\cosh x$	$\sinh x$
$\tanh x$	$\mathrm{sech}^2\, x$
$\sinh^{-1} x$	$\dfrac{1}{\sqrt{(x^2 + 1)}}$
$\cosh^{-1} x$	$\dfrac{1}{\sqrt{(x^2 - 1)}}$
$\tanh^{-1} x$	$\dfrac{1}{1 - x^2}$

Examples

1 Differentiate w.r.t. to x

(i) $\sinh 3x$ (ii) $\cosh(8x^3 + 2x + 1)$

(i) $\dfrac{d}{dx}\sinh 3x = 3\cosh 3x$

(ii) $\dfrac{d}{dx}\cosh(8x^3 + 2x + 1) = \sinh(8x^2 + 2x + 1).(24x^2 + 2)$

$$= 2(12x^2 + 1)\sinh(8x^3 + 2x + 1)$$

2 Differentiate w.r. to x

(i) $\sinh^{-1} 3x$ (ii) $\tanh^{-1} 4x^2$

(i) $$\frac{d}{dx}\sinh^{-1} 3x = \frac{1}{\sqrt{(9x^2+1)}}.3 = \frac{3}{\sqrt{(9x^2+1)}}$$

(ii) $$\frac{d}{dx}\tanh^{-1} 4x^2 = \frac{1}{1-16x^4}.8x = \frac{8x}{1-16x^4}$$

Practice questions 6

1 Differentiate w.r. to x

(i) $\sinh 9x$ (ii) $\cosh x^2$ (iii) $\tanh 8x$

2 Differentiate w.r. to x

(i) $\sinh^{-1} 8x$ (ii) $\cosh^{-1} 5x^2$ (iii) $\tanh^{-1} 3x^2$

Logarithmic differentiation

When complicated products or quotients or functions of the form $[f(x)]^{q(x)}$ have to be differentiated the work is simplified if the logarithms of the given function is taken before differentiating. The process is illustrated in the following examples.

Examples

1 Find the differential coefficient w.r. to x of

$$\frac{x^2(3+x^2)^4}{\sqrt{(1-x^3)}}$$

$$y = \frac{x^2(3+x^2)^4}{\sqrt{(1-x^3)}}$$

Take the logarithms to base e of both sides

$$\log y = \log\frac{x^2(3+x^2)^4}{(1-x^3)^{1/2}}$$

$$\log y = \log x^2 + \log(3+x^2)^4 - \log(1-x^3)^{1/2}$$

$$\log y = 2\log x + 4\log(3+x^2) - \tfrac{1}{2}\log(1-x^3)$$

differentiating w.r. to x

$$\frac{1}{y}\frac{dy}{dx} = \frac{2}{x} + \frac{4}{3+x^2}.2x - \frac{1}{2}.\frac{1}{1-x^3}.(-3x^2)$$

$$\frac{dy}{dx} = y\left[\frac{2}{x} + \frac{8x}{3+x^2} + \frac{3x^2}{2(1-x^3)}\right]$$

$$= \frac{x^2(3+x^2)^4}{\sqrt{(1-x^3)}}\left[\frac{2}{x} + \frac{8x}{3+x^2}\ \frac{3x^2}{2(1-x^2)}\right]$$

2 Differentiate w.r. to x

$$(\sin 2x)^{x^2}$$

$$y = (\sin 2x)^{x^2}$$

$$\log y = x^2 \log(\sin 2x)$$

$$\frac{1}{y}\frac{dy}{dx} = 2x\log(\sin 2x) + x^2 \cdot \frac{1}{\sin 2x} \cdot \cos 2x \cdot 2$$

$$\frac{dy}{dx} = y\left[2x\log(\sin 2x) + \frac{2x^2\cos 2x}{\sin 2x}\right]$$

$$\frac{dy}{dx} = 2x(\sin 2x)^{x^2}[\log(\sin 2x) + x\cot 2x]$$

Practice questions 7

1 Differentiate w.r. to x

(i) $(x^2+3x-1)^3(x-4)^5$

(ii) $\sqrt{x^2+5}\,(3x^4-x^2-1)^3$

2 Differentiate w.r. to x

(i) $(\tan 6x)^x$ (ii) $(3x^2-x-1)^{x^2}$

3 Differentiate w.r. to x

(i) $\dfrac{(3x+1)^3(5x+4)^2}{(2-x)^4}$

(ii) $\dfrac{(8-3x)^{1/2}(7-x)^{3/4}}{\sinh x}$

Successive differentiation

If y is a function of x then in general $\frac{dy}{dx}$ is a function of x. The derivative of $\frac{dy}{dx}$ i.e. $\frac{d}{dx}\left(\frac{dy}{dx}\right)$ is denoted by $\frac{d^2y}{dx^2}$. Its derivative is denoted by $\frac{d^3y}{dx^3}$ and so on. Alternative notations are $f'(x), f''(x), f'''(x)$ and y_1, y_2, y_3 for the first three derivatives. Sometimes a general formula for the nth derivative of a function can be found.

Examples

1 Find the first four derivatives of the function ax^p, a a constant, and hence find a formula for the nth derivative of this function.

$$y = ax^p$$
$$y_1 = apx^{p-1}$$
$$y_2 = ap(p-1)x^{p-2}$$
$$y_3 = ap(p-1)(p-2)x^{p-3}$$
$$y_4 = ap(p-1)(p-2)(p-3)x^{p-4}$$
$$\cdots\cdots\cdots\cdots$$

in general.

$$y_n = ap(p-1)(p-2)(p-3)\ldots[p-(n-1)]x^{p-n}$$
$$\mathbf{y_n = ap(p-1)(p-2)\ldots(p-n+1)x^{p-n}}$$

2 Find a formula for the nth differential coefficient of the function $\sin bx$, (b is a constant), w.r. to x

$$y = \sin bx$$
$$y_1 = b\cos bx$$
$$y_2 = -b^2 \sin bx$$
$$y_3 = -b^3 \cos bx$$
$$y_4 = b^4 \sin bx$$

the pattern of the trigonometric functions is then repeated

Now $$\cos bx = \sin\left(bx + \frac{\pi}{2}\right)$$
$$-\sin bx = \sin(bx + \pi)$$
and $$-\cos bx = \sin\left(bx + \frac{3\pi}{2}\right)$$

Hence

$$y = \sin bx$$
$$y_1 = b\sin\left(bx + \frac{\pi}{2}\right)$$
$$y_2 = b^2 \sin(bx + \pi)$$
$$y_3 = b^3 \sin\left(bx + \frac{3\pi}{2}\right)$$
$$y_4 = b^4 \sin(bx + 2\pi) = b^4 \sin\left(bx + \frac{4\pi}{2}\right)$$

in general, $\mathbf{y_n = b^n \sin\left(bx + \frac{n\pi}{2}\right)}$

Practice questions 8

1 Find the first four derivatives of the function $(ax + b)^m$ and hence find a formula for the nth derivative of this function (w.r. to x).

2 Find a formula for the nth differential coefficient will respect to x of

(i) $\sin(ax + b)$ (ii) $\cos ax$

3 Find the fifth derivative of $e^x \cos x$ w.r. to x.
Deduce formula for the nth derivative.

Differentiation of parametric equations

x and y are both given in terms of a third variable, the parameter, e.g.

$$x = f(t) \quad \text{and} \quad y = g(t)$$

then

$$\frac{dx}{dt} = f(t) \quad \text{and} \quad \frac{dy}{dt} = g'(t)$$

$$\frac{dy}{dx} = \frac{\frac{dy}{dt}}{\frac{dx}{dt}}$$

$$\frac{dy}{dx} = \frac{g'(t)}{f'(t)}$$

So $\frac{dy}{dx}$ is itself a function of t and equals $\phi(t)$ say.

$$\frac{dy}{dx} = \phi(t)$$

$$\frac{d^2y}{dx^2} = \frac{d}{dx}\left(\frac{dy}{dx}\right)$$

$$= \frac{d}{dx}[\phi(t)]$$

$$= \frac{d}{dt}[\phi(t)] \cdot \frac{dt}{dx}$$

$$\frac{d^2y}{dx^2} = \frac{\frac{d}{dt}[\phi(t)]}{\frac{dx}{dt}}$$

Both derivatives are functions of t.
Higher derivatives may be obtained using the same method.

Example

Find $\frac{dy}{dx}$ and $\frac{d^2y}{dx^2}$ when $x = 2(\theta - \sin\theta)$ and $y = 2(1 - \cos\theta)$

$x = 2(\theta - \theta \sin)$, $\quad y = 2(1 - \cos\theta)$

$\frac{dx}{d\theta} = 2 - 2\cos\theta \qquad \frac{dy}{d\theta} = 2\sin\theta$

$$\frac{dy}{dx} = \frac{\frac{dy}{d\theta}}{\frac{dx}{d\theta}} = \frac{2\sin\theta}{2(1-\cos\theta)}$$

$$= \frac{2\sin\frac{\theta}{2}\cos\frac{\theta}{2}}{2\sin^2\frac{\theta}{2}}$$

$$\frac{dy}{dx} = \cot\frac{\theta}{2}$$

$$\frac{d^2y}{dx} = \frac{d}{dx}\left(\frac{dy}{dx}\right)$$

$$= \frac{d}{dx}\left(\cot\frac{\theta}{2}\right)$$

$$= \frac{d}{d\theta}\left(\cot\frac{\theta}{2}\right).\frac{d\theta}{dx}$$

$$= -\operatorname{cosec}^2\frac{\theta}{2}.\frac{1}{2}.\frac{1}{\frac{dx}{d\theta}}$$

$$= -\tfrac{1}{2}\operatorname{cosec}^2\frac{\theta}{2}.\frac{1}{2(1-\cos\theta)}$$

$$= -\tfrac{1}{4}\operatorname{cosec}^2\frac{\theta}{2}.\frac{1}{2\sin^2\frac{\theta}{2}}$$

$$\frac{d^2y}{dx} = -\tfrac{1}{8}\operatorname{cosec}^4\frac{\theta}{2}$$

Practice questions 9

1 Find $\frac{dy}{dx}$ and $\frac{d^2y}{dx^2}$ when $x = 4\cos^3\theta$ and $y = 4\sin^3\theta$

2 Find $\frac{dy}{dx}$ and $\frac{d^2y}{dx^2}$ in terms of t when

$$x = \frac{1-t^2}{1+t^2} \text{ and } y = \frac{2t}{1+t^2}$$

3 Find $\frac{dy}{dx}$ and $\frac{d^2y}{dx^2}$ in terms of θ when

$$x = \cos\theta + \theta\sin\theta \quad \text{and } y = \sin\theta - \theta\cos\theta$$

Differentiation of implicit functions

An implicit function is an equation in x and y in which neither variable can be conveniently expressed in terms of the other. The equation is differentiated term by term with respect to x giving an equation in dy/dx, x and y, which is then solved for dy/dx. This equation is differentiated term by term with respect to x giving an equation in d^2y/dx^2, dy/dx x and y which is then solved for d^2y/dx^2 etc.

Example
Find dy/dx and d^2y/dx^2 when $x^2 + 2xy - y^2 = 5$

$$x^2 + 2xy - y^2 = 5$$

differentiating term by term w.r. to x

$$2x + 2\left(1 . y + x\frac{dy}{dx}\right) - 2y\frac{dy}{dx} = 0$$

$$x + y + x\frac{dy}{dx} - y\frac{dy}{dx} = 0$$

$$x + y = \frac{dy}{dx}(y - x)$$

$$\frac{dy}{dx} = \frac{x+y}{y-x} \qquad \text{(i)}$$

differentiating (i) w.r. to x

$$\frac{d^2y}{dx^2} = \frac{(y-x)\left(1 + \frac{dy}{dx}\right) - (x+y)\left(\frac{dy}{dx} - 1\right)}{(y-x)^2}$$

$$= \frac{y - x + x + y + \frac{dy}{dx}(y - x - x - y)}{(y-x)^2}$$

$$= \frac{2y + \frac{dy}{dx}(-2x)}{(y-x)^2}$$

$$= 2\frac{y + \frac{x+y}{y-x}(-x)}{(y-x)^2} \qquad \text{using (i)}$$

$$= 2\frac{y(y-x) - x(x+y)}{(y-x)^3}$$

$$= 2\frac{y^2 - xy - x^2 - xy}{(y-x)^3}$$

$$= 2\frac{y^2 - 2xy - x^2}{(y-x)^3}$$

But $x^2 + 2xy - y^2 = 5$, given

$$\therefore \ \frac{d^2y}{dx^2} = \frac{-10}{(y-x)^3}$$

Practice questions 10

1 Find $\frac{dy}{dx}$ when $x^3 + 3x^2y + y^3 = 4$

2 Find $\frac{dy}{dx}$ when $y = \sin(x + y)$

3 Find $\frac{dy}{dx}$ and $\frac{d^2y}{dx^2}$ when $x^2 + 2xy - 4y^2 = 6$

Application of differentiation

The gradient of a curve

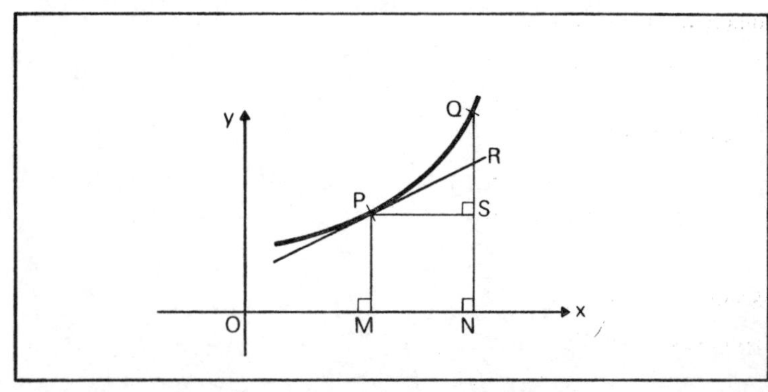

$P(x, y)$ and $Q(x + \delta x, y + \delta y)$ are neighbouring points on a continuous curve $y = f(x)$.
The tangent to the curve at P is defined as the limiting position of the chord PQ as Q moves along the curve towards P. The gradient of the curve at P is defined as the gradient of the tangent at P. In the above diagram $PS = \delta x$, $SQ = \delta y$

$$SQ = f(x + \delta x) - f(x)$$

$$\text{gradient of the chord } PQ = \frac{SQ}{PS}$$

$$= \frac{\delta y}{\delta x}$$

$$= \frac{f(x + \delta x) - f(x)}{\delta x}$$

$$\text{gradient of curve at } P \text{ is } \lim_{\delta x \to 0} \frac{\delta y}{\delta x} = \lim_{\delta x \to 0} \frac{f(x + \delta x) - f(x)}{\delta x}$$

$$= \frac{dy}{dx}$$

$$= f'(x)$$

Example
Find the equations of the tangent and normal to the curve $y = 4x^4 + 2x^2 + 1$ at the point where $x = \frac{1}{2}$

$$y = 4x^4 + 2x^2 + 1 \qquad \text{(i)}$$

when $x = \frac{1}{2}$,

$$y = 4 \cdot \tfrac{1}{16} + 2 \cdot \tfrac{1}{4} + 1$$

$$= \tfrac{1}{4} + \tfrac{1}{2} + 1 = 1\tfrac{3}{4}$$

differentiating (i) w.r. to x

$$\frac{dy}{dx} = 16x^3 + 4x$$

when $x = \frac{1}{2}$,

$$\frac{dy}{dx} = 16 \cdot \tfrac{1}{8} + 4 \cdot \tfrac{1}{2}$$

$$= 2 + 2 = 4$$

gradient of tangent at $(\frac{1}{2}, 1\frac{3}{4})$ is 4

equation of tangent at $(\frac{1}{2}, 1\frac{3}{4})$ is

$$y - 1\tfrac{3}{4} = 4(x - \tfrac{1}{2})$$

$$y - 1\tfrac{3}{4} = 4x - 2$$

$$y = 4x - \tfrac{1}{4} \text{ i.e. } 16x - 4y - 1 = 0$$

gradient of normal at $(\frac{1}{2}, 1\frac{3}{4})$ is $-\frac{1}{4}$

equation of normal at $(\frac{1}{2}, 1\frac{3}{4})$ is

$$y - 1\tfrac{3}{4} = -\tfrac{1}{4}(x - \tfrac{1}{2})$$

$$4y - 7 = -x + \tfrac{1}{2}$$

$$8y - 14 = -2x + 1$$

$$2x + 8y - 15 = 0$$

Practice questions 11

1 Find the equations of the tangent and normal to the curve $y = 2x^3 - 9x + 4$ at the point whose x-coordinate is 2.

2 Find the coordinates of the point at which the tangent to the curve $y = x^2 - 5x - 3 \ln x$ is parallel to the x-axis.

Maximum, minimum and points of inflexion

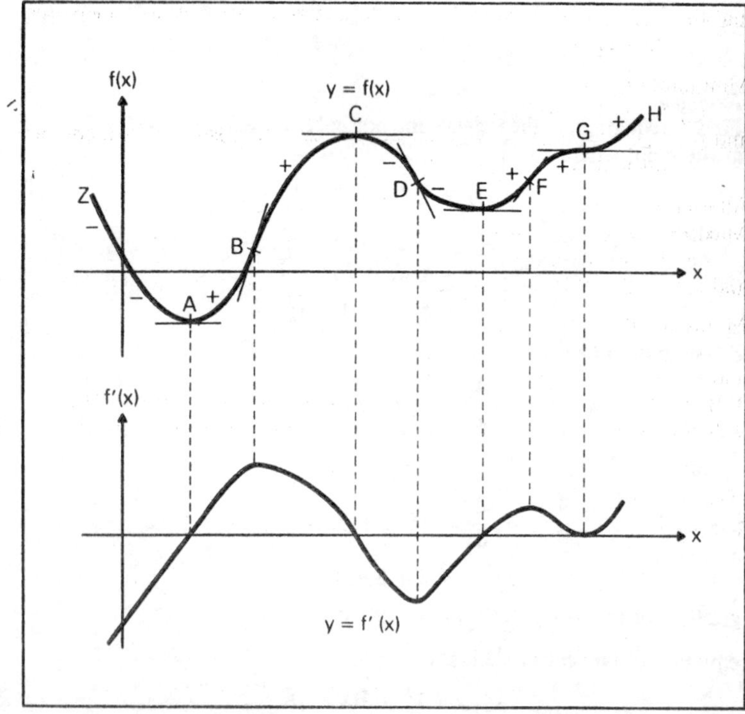

The first graph is that of some differentiable function $f(x)$ and the second graph is that of the corresponding function $f'(x)$. The positive or negative nature of $f(x)$ is indicated by + and − signs on the graph of $f(x)$.

Points A, C, E and G are called **stationary** points, i.e. points on $y = f(x)$ where $f'(x) = 0$.

Points A, C and E are called **turning** points i.e. points where $f'(x) = 0$ and also $f'(x)$ changes sign as x increases through the point.

A and E are **minimum** points i.e. points where $f(x)$ has a local minimum value.

C is a **maximum** point i.e. a point where $f(x)$ has a local maximum value.

G is a horizontal point of inflexion.

B, D and F are general points of inflexion.

G, B, D and F give corresponding turning points on the curve $y = f'(x)$. i.e. points where $f''(x) = 0$ and $f''(x)$ changes sign as x increases through the points.

Tests for maximum and minimum values of $f(x)$
($f(x)$ continuous at $x = x'$)

Maximum at $x = x_1$ when
(a) $f'(x) = 0$ when $x = x_1$
and (b) $f'(x)$ changes sign from positive to zero to negative as x increases through x_1

Minimum at $x = x_1$ when
(a) $f'(x) = 0$ when $x = x_1$
and (b) $f'(x)$ changes sign from negative to zero to positive as x increases through x_1.

Alternatively,
Maximum at $x = x_1$ when
(a) $f'(x) = 0$ when $x = x_1$
and (b) $f''(x) < 0$ when $x = x_1$

Minimum at $x = x_1$ when
(a) $f'(x) = 0$ when $x = x_1$
and (b) $f''(x) > 0$ when $x = x_1$

N.B. If when using the alternative tests you obtain $f'(x_1) = 0$ and $f''(x_1) = 0$ then no conclusion can be drawn and the other tests should be used.

Concavity of a curve

The graph of a differentiable function $f(x)$ is said to be concave upward on an interval if $f'(x)$ is an increasing function on the interval; it is said to be concave downward if $f'(x)$ is decreasing on the interval. A point where the concavity of a curve changes is said to be a point of inflexion (inflection). The graph of $y = f(x)$ above is concave upward over ZAB, DEF and GH.

If $f''(x) > 0$ in an interval, $f(x)$ is concave upward in the interval
If $f''(x) < 0$ in an internal, fx is concave downward in the interval

Tests for a point of inflexion at $x = x_1$

(a) $f''(x) = 0$ when $x = x_1$

and (b) $f''(x)$ changes sign as x increases through $x = x_1$

alternatively

(a) $f''(x) = 0$ when $x = x_1$

and (b) $f'''(x) \neq 0$ when $x = x_1$

N.B. If $f''(x) = 0$ and $f'''(x) = 0$ then no conclusion can be drawn and the other test should be used.

Example

1 Find the local maxima, minima and points of inflexion of the function

$$\frac{16x}{x^2+4}$$

$$f(x) = \frac{16x}{x^2+4}$$

$$f'(x) = 16.\frac{(x^2+4)1 - x.2x}{(x^2+4)^2}$$

$$= 16.\frac{4-x^2}{(x^2+4)^2}$$

$$f''(x) = 16.\frac{(x^2+4)^2.(-2x) - (4-x^2).2(x^2+4).2x}{(x^2+4)^4}$$

$$= -32x\frac{x^2+4+8-2x^2}{(x^2+4)^3}$$

$$= -32.\frac{x(12-x^2)}{(x^2+4)^3}$$

At stationary points $f'(x) = 0$

i.e. $$\frac{4-x^2}{(x^2+4)^2} = 0$$

$$x^2 = 4$$

$$x = \pm 2$$

when $x = -2$, $f''(x) = \dfrac{-32(-2)(8)}{(8)^3} > 0$

$\therefore$ **minimum when $x = -2$ and $f(-2) = -4$**

when $x = 2$, $f''(x) = \dfrac{-32.(2)(8)}{(8)^3} < 0$

$\therefore$ **maximum when $x = 2$, and $f(2) = 4$**

At points of inflexion, (a) (a) $f''(x) = 0$

i.e. $$\frac{x(12 - x^2)}{(x^2 + 4)^3} = 0$$

$$x(12 - x^2) = 0$$

$$x = -\sqrt{12},\ 0,\ \sqrt{12}$$

i.e. $$x = -2\sqrt{3}.0,\ 2\sqrt{3}$$

in neighbourhood of $x = -2\sqrt{3}$,

when $x < -2\sqrt{3}$, $f''(x) > 0$

when $x > -2\sqrt{3}$, $f''(x) < 0$

(b) $f''(x)$ changes sign

$\therefore$ **there is a point of inflexion when x = $-2\sqrt{3}$ and f($-2\sqrt{3}$) = $-2\sqrt{3}$.**

in neighbourhood of $x = 0$

when $x < 0$, $f''(x) > 0$

when $x > 0$, $f''(x) < 0$

(b) $f''(x)$ changes sign

$\therefore$ **there is a point of inflexion when x = 0 and f(0) = 0**

in neighbourhood of $x = 2\sqrt{3}$

when $x < 2\sqrt{3}$, $f''(x) < 0$

when $x > 2\sqrt{3}$, $f''(x) > 0$

(b) $f''(x)$ changes sign

there is a point of inflexion when x = $2\sqrt{3}$ and f($2\sqrt{3}$) = $2\sqrt{3}$

Practice questions 12

1 Find the local maxima and minima of the function

$$3x^4 + 8x^3 - 12x^2 + 12.$$

2 Find and investigate the nature of the stationary points and the points of inflexion on the curve $y = xe^{-2x^2}$.

3 Find and identify the turning point of the curve

$$y = \frac{e^{x/3}}{e^x + 1}$$

4 Find and investigate the nature of the stationary points on the curve

$$y = 4\sqrt{3}\cos x - 2\cos 2x,$$

for values of x from $-\frac{\pi}{4}$ to $+\frac{9\pi}{4}$

5 Find the turning points and points of inflexion on the curve

$y = (x - 4)(x + 1)^3$. Sketch the curve

Curvature

Radius of curvature

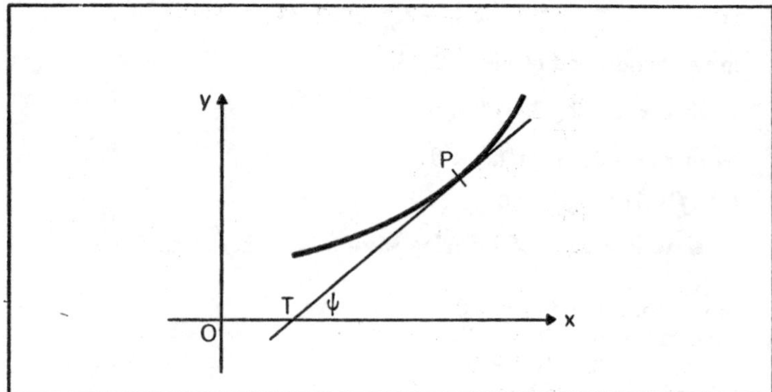

The tangent, PT, at P to the curve $y = f(x)$ makes an angle ψ with Ox where

$$\tan\psi = \frac{dy}{dx} = f'(x)$$

The curvature of the curve is the rate of change of ψ with respect to s, the arc length.
The **curvature of the curve at P** is $d\psi/ds$.
A circle touching the curve at P, with its concavity in the same direction and having the same curvature as the curve at P is called the **circle of curvature at P.** Its radius ρ is called the **radius of curvature at P** and its centre is called the **centre of curvature at P.**

$$\rho = \frac{1}{\dfrac{d\psi}{ds}} = \frac{ds}{d\psi}$$

Formula for ρ when the equation of the curve is in the form $y = f(x)$

$$\rho = \frac{\left[1 + \left(\dfrac{dy}{dx}\right)^2\right]}{\dfrac{d^2y}{dx^2}}$$

Convention

(i) The numerator is taken as positive.

(ii) Then ρ has the same sign as $\dfrac{d^2y}{dx^2}$ and is therefore positive when the curve is concave upwards and negative when the curve is concave downwards.

Examples

1 Find the radius of curvature of the curve.

$$y = 12 + 3x^2 + x^3 \text{ at the point } (1, 16)$$

$$y = 12 + 3x^2 + x^2$$

$$\frac{dy}{dx} = 6x + 3x^2$$

$$= 3(2x + x^2)$$

$$\frac{d^2y}{dx} = 3(2 + 2x)$$

$$= 6(1 + x)$$

$$\rho = \frac{\left[1 + \left(\frac{dy}{dx}\right)^2\right]^{3/2}}{\frac{d^2y}{dx^2}}$$

$$= \frac{[1 + 9(2x + x^2)^2]^{3/2}}{6(1 + x)}$$

At (1, 16) $$\rho = \frac{(1 + 81)^{3/2}}{12} = \frac{(82)^{3/2}}{12}$$

2 Find the radius of curvature of the cycloid

$x = \theta + \sin\theta, \quad y = 1 + \cos\theta$

at the point where $\theta = \frac{\pi}{3}$

$$x = \theta + \sin\theta \qquad y = 1 + \cos\theta$$

$$\frac{dx}{d\theta} = 1 + \cos\theta \qquad \frac{dy}{d\theta} = -\sin\theta$$

$$\frac{dy}{dx} = \frac{\frac{dy}{d\theta}}{\frac{dx}{d\theta}} = \frac{-\sin\theta}{1 + \cos\theta} = -\frac{2\sin\frac{\theta}{2}\cos\frac{\theta}{2}}{2\cos^2\frac{\theta}{2}}$$

$$= -\tan\frac{\theta}{2}$$

$$\frac{d^2y}{dx^2} = \frac{d}{dx}\left(\frac{dy}{dx}\right) = \frac{d}{d\theta}\left(\frac{dy}{dx}\right)\frac{d\theta}{dx}$$

$$= \frac{\frac{d}{d\theta}\left(-\tan\frac{\theta}{2}\right)}{\frac{dx}{d\theta}}$$

$$= \frac{-\frac{1}{2}\sec^2\frac{\theta}{2}}{1+\cos\theta} = \frac{-\frac{1}{2}\sec^2\frac{\theta}{2}}{2\cos^2\frac{\theta}{2}}$$

$$= -\tfrac{1}{4}\sec^4\frac{\theta}{2}$$

$$\rho = \frac{\left[1+\left(\frac{dy}{dx}\right)^2\right]^{3/2}}{\frac{d^2y}{dx^2}}$$

$$= \frac{\left[1+\tan^2\frac{\theta}{2}\right]}{-\frac{1}{4}\sec^4\frac{\theta}{2}}$$

$$= -4.\frac{\sec^3\frac{\theta}{2}}{\sec^4\frac{\theta}{2}} = -4\frac{1}{\sec\frac{\theta}{2}}$$

$$= -4\cos\frac{\theta}{2}$$

When $\theta = \frac{\pi}{3}$, $\rho = -4\cos\frac{\pi}{6} = -2\sqrt{3}$

the radius of curvature at $\theta = \pi/3$ is $2\sqrt{3}$ units

Practice questions 13

1 Find the radii of curvature of the following curves at the stated points

(a) $y = \sin x$ at $x = \frac{\pi}{4}$

(b) $x = 2(\theta - \sin\theta)$, $y = 2(1 - \cos t)$ at $x = \dfrac{\pi}{2}$

2 For the catenary $y = c\cosh x/c$ prove that $\rho = y^2/c$.
3 Find the radius of curvature at the point with parameter t on the curve

$$x = 2\cos^3 t \qquad y = 2\sin^3 t.$$

Curve sketching

(a) *Cartesian co-ordinates*

1 Inspect the equation of the curve for an obvious change of origin which will give a simplified equation.
e.g. $(y + 3)^4 = (x - 2)^3$ reduces to $y^4 = x^3$ when the origin is moved to the point $(2, -3)$.
In general the equation $f(x + a, y + b) = 0$ reduces to $f(X, Y)$ if the origin is moved to $(-a, -b)$ and $X = x + a$, $Y = y + b$.

2 Inspect the equation for symmetry. If there are no odd powers of x the curve is symmetrical about the y-axis; if there are no odd powers of y the curve is symmetrical about the x-axis, e.g. $y^4 = x^3$ above is symmetrical about the new x-axis.

3 Determine where the curve cuts the axes.

4 Determine any obvious points on the curve.

5 Determine whether there are any restrictions on the ranges of x and/or y. These may be found by expressing y in terms of x, or x in terms of y.
e.g. $y^2 = x(x^2 - 4)$, y is real only when $x \geq 4$ and when $-2 \leq x \leq 0$.

6 Determine any asymptotes parallel to the axes

e.g. $y = \dfrac{1}{x - 1}$, $x = 1$ is an asymptote.

7 Investigate the concavity of the curve.

8 Find maximum and minimum points, if any.

9 Find any points of inflexion.

10 Examine $\lim\limits_{x \to +\infty} f(x)$ and $\lim\limits_{x \to -\infty} f(x)$.

11 Determine any general asymptotes.
When sketching a particular curve you choose an appropriate selection of tests from numbers 1 to 11 above in order to determine the shape of the curve.

Examples

1 Sketch the curve given by the equation $y = x^2(x - 2)^2$.

Test 2, No symmetry about either axes.
Test 3, $y = 0$ when $x = 0, 2$
when $x = 0$ $y = 0$
Test 5, since $x^2(x - 2)^2$ is a perfect square $y \geq 0$
Test (7) 8 (9)

$$y = x^2(x - 2)^2$$

$$\frac{dy}{dx} = 2x(x - 2)^2 + 2x^2(x - 2)$$

$$= 2x(x - 2)(x - 2 + x)$$

$$= 4x(x - 2)(x - 1)$$

$$= 4(x^3 - 3x^2 + 2x)$$

$$\frac{d^2y}{dx^2} = 4(3x^2 - 6x + 2)$$

At turning points, $\frac{dy}{dx} = 0$

$$x(x - 2)(x - 1) = 0$$

$$x = 0, 1, 2$$

When $x = 0$, $\frac{d^2y}{dx^2} = 8$, i.e. a minimum point at $x = 0$, $y = 0$

When $x = 1$, $\frac{d^2y}{dx^2} = -4$, i.e. a maximum point at $x = 1$, $y = 1$

When $x = 2$, $\frac{d^2y}{dx^2} = 8$, i.e. a minimum point at $x = 2$, $y = 0$

Since $\frac{d^2y}{dx^2} > 0$ for large x (numerically) curve is then concave upwards.

Test 10 As $x \to +\infty$ $y \to +\infty$
As $x \to -\infty$ $y \to +\infty$

We are now in a position to outline a sketch of the curve. For a more detailed sketch the points of inflection should be determined.

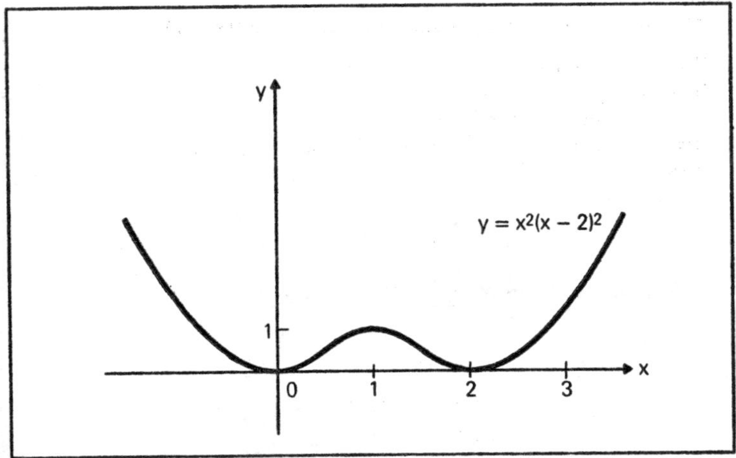

2 Find turning points and points of inflexion on the curve

$$y = \frac{8(x-3)}{(x-4)^2}.\ \text{Sketch the curve.}$$

$$y = 8 \, . \frac{x-3}{(x-4)^2}$$

$$\frac{dy}{dx} = 8 \, . \frac{(x-4)^2 1 - (x-3)2(x-4)1}{(x-4)^4}$$

$$= 8 \, . \frac{x-4-2(x-3)}{(x-4)^3},$$

$$= 8 \, . \frac{-x+2}{(x-4)^3}$$

At turning points $\frac{dy}{dx} = 0, \quad -x+2=0$

$$x = 2$$

In the neighbourhood of $x = 2$, when $x < 2$ $\frac{dy}{dx}$ is $-$ve

and when $x > 2$ $\frac{dy}{dx}$ is $+$ve

$\therefore$ **there is a minimum point at (2, −2)**

$$\frac{d^2y}{dx^2} = 8 \, . \frac{(x-4)^3(-1) - (2-x)3(x-4)^2}{(x-4)^6}$$

$$= 8 . \frac{-(x-4) - 3(2-x)}{(x-4)^4}$$

$$8 . \frac{2x-2}{(x-4)^4}$$

$$= 4 . \frac{x-1}{(x-4)^4}$$

At points of inflexion $\frac{d^2y}{dx} = 0, \quad x - 1 = 0$

$$x = 1$$

In neighbourhood of $x = 1$, when $x < 1 \frac{d^2y}{dx^2}$ is $-$ve

and when $x > 1 \frac{d^2y}{dx^2}$ is $+$ve

$\therefore$ **there is a point of inflexion at** $\left(1, -\frac{16}{9}\right)$

When $x = 4$, y is infinite
$\therefore$ there is an asymptote parallel to the y axis with equation $x = 4$.
When $x = 0$ $y = -\frac{3}{2}$
When $y = 0$ $x = 3$

When $x < 1 \frac{d^2y}{dx^2} = 4 \quad \frac{x-1}{(x-4)^4} < 0$ curve is concave downwards.

When $1 < x < 4 \frac{d^2y}{dx^2} > 0$ curve is concave upwards.

When $x > 4 \frac{d^2y}{dx^2} > 0$ curve is concave upwards.

As $x \to 4$ from below $y = \frac{8(x-3)}{(x-4)^2} \to +\infty$

As $x \to 4$ from above $y \to +\infty$

As $x \to +\infty$ $y \to 0$ from above

As $x \to -\infty$ $y \to 0$ from below
The x-axis is an asymptote.

N.B. There is bound to be distortion of the curve on the vertical scale with a curve of this type.

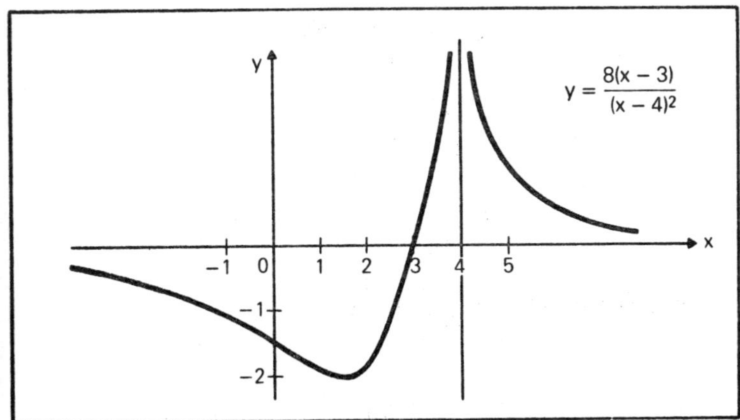

3 Find the turning points on the curve

$y = \frac{9x^2 + 6x + 51}{5(2x - 1)}$. **Sketch the curve**

$$y = \frac{9x^2 + 6x + 51}{5(2x - 1)}$$

$$\frac{dy}{dx} = \frac{(10x - 5)(18x + 6) - (9x^2 + 6x + 51)10}{25(2x - 1)^2}$$

$$= \frac{180x^2 - 30x - 30 - 90x^2 - 60x - 510}{25(2x - 1)^2}$$

$$= \frac{90x^2 - 90x - 540}{25(2x - 1)^2}$$

$$= 90 . \frac{x^2 - x - 6}{25(2x - 1)^2}$$

$$90 . \frac{(x - 3)(x + 2)}{25(2x - 1)^2}$$

At turning points $\frac{dy}{dx} = 0, \quad (x - 3)(x + 2) = 0$

$$x = -2, 3$$

In neighbourhood of $x = -2$, when $x < -2$, $\frac{dy}{dx}$ is +ve and when $x > -2 \dfrac{dy}{dx}$ is $-$ve

$\therefore$ there is a maximum point at $x = -2$

when $x = -2$,

$$y = \frac{9(-2)^2 + 6(-2) + 51}{5(-4-1)}$$

$$= \frac{36 - 12 + 51}{-25}$$

$$= \frac{75}{-25}$$

$$= -3$$

$\therefore$ **maximum point at $(-2, -3)$**

In neighbourhood of $x = 3$, when $x < 3 \dfrac{dy}{dx}$ is $-$ve

and when $x > 3 \dfrac{dy}{dx}$ is +ve

$\therefore$ there is a minimum point at $x = 3$

when $x = 3$,

$$y = \frac{9.3^2 + 18 + 51}{5(6-1)}$$

$$= \frac{150}{25}$$

$$= 6$$

$\therefore$ **minimum point at (3, 6)**

As $x \to +\infty \quad y \to +\infty$
As $x \to -\infty \quad y \to -\infty$

y is infinite also when $2x - 1 = 0$

$$x = \tfrac{1}{2}$$

the line $x = \frac{1}{2}$ is an asymptote

As $x \to \frac{1}{2}$ from below $y \to -\infty$

As $x \to \frac{1}{2}$ from above $y \to +\infty$

when $x = 0 \quad y = -\dfrac{51}{5}$

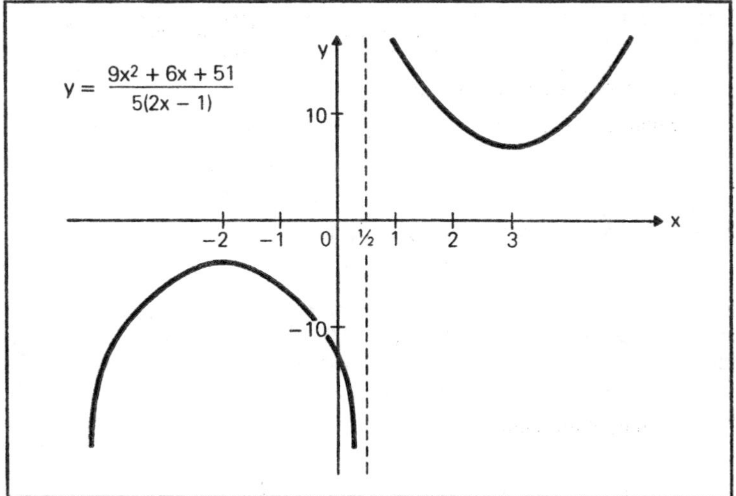

Practice questions 14

1 Sketch the following curves

(i) $y = \dfrac{20}{x^2 + 1}$ (ii) $y = \dfrac{(x-1)(x-3)}{x(x+2)}$

2 Find the turning points and points of inflexion on the curve $y = x^3 - 8x^2 - 12x + 5$. Sketch the curve.

3 Sketch the curve $y = \dfrac{1}{1 - x^2}$.

4 Find the turning points and points of inflexion on the curve

$$y = \frac{3x}{x^2 + 1}.$$

Sketch the curve.

(b) *Polar coordinates*
The equation of the curve is given in the form $r = f(\theta)$.
If $f(-\theta) = f(\theta)$ the curve is symmetrical about the line $\theta = 0$.
If $f(-\theta)$ $-f(\theta)$ the curve is symmetrical about the line $\theta = \pi/2$.
If $f(\pi - \theta) = f(\theta)$ the curve is symmetrical about the line $\theta = \pi/2$.
In particular if r is a function of $\cos\theta$ only the curve is symmetrical about the initial line and if r is a function of $\sin\theta$ only the curve is symmetrical about the line $\theta = \pi/2$.

If the equation of the curve is such that it contains only even powers of r there is symmetry about the pole (origin).
An examination of the equation of the curve will show if there are any limitations on the values of r and θ.
Before sketching the curve it is usually useful to tabulate some values of r for certain θ.

Example
Sketch the curve given by the polar equation

$$r = 4\cos 2\theta$$

Since $\cos 2\theta = 2\cos^2\theta - 1$, a function of $\cos\theta$ only, the curve is symmetrical about the initial line, $\theta = 0$. Also $\cos 2\theta = 1 - 2\sin^2\theta$, a function of $\sin\theta$ only, the curve is symmetrical also about the line $\theta = \pi/2$. Consequently we need only values of r corresponding to values of θ from 0 to $\pi/2$.

$r = 4\cos 2\theta$
and
$|\cos 2\theta| \leq 1$
so
$|r| \leq 4$

Construct the following table

$\theta = 0$	$\frac{\pi}{12}$	$\frac{\pi}{8}$	$\frac{\pi}{6}$	$\frac{\pi}{4}$	$\frac{\pi}{3}$	$\frac{3\pi}{8}$	$\frac{5\pi}{12}$	$\frac{\pi}{2}$
$r = 4$	$2\sqrt{3}$	$2\sqrt{2}$	2	0	-2	$-2\sqrt{2}$	$-2\sqrt{3}$	-4

We thus obtain:

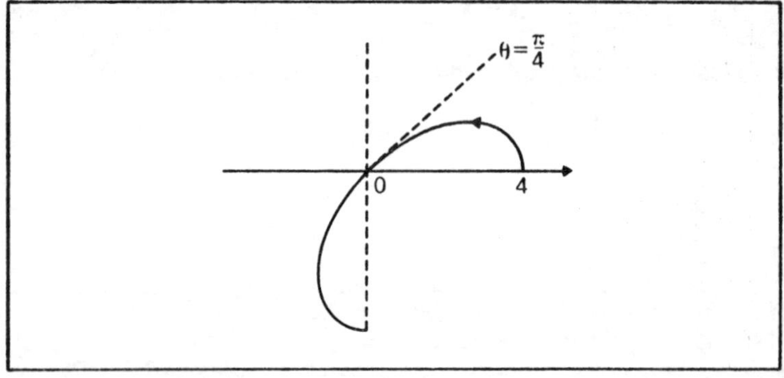

Using the symmetrical properties we can now complete the curve which is 4-leaved rose.

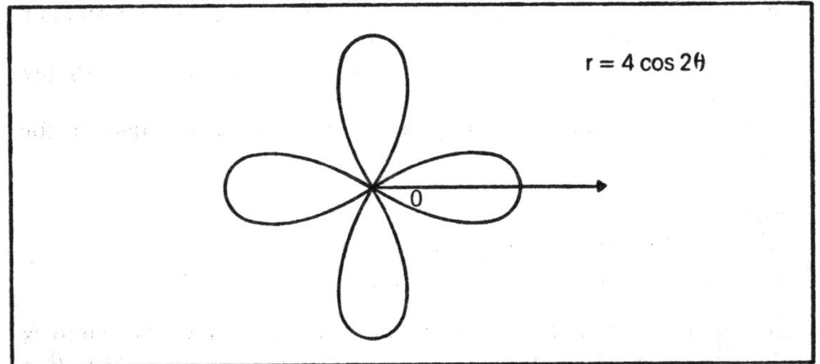

Practice questions 15

Sketch and name the curves given by the following polar equations.

1 $r = 2 + 3\cos\theta$ 2 $r^2 = 9\cos 2\theta$
3 $r = 10\sin 3\theta$ 4 $r = 6(1 + \cos\theta)$

Integration

Standard integrals, the arbitrary constant is **omitted** in each case.

$$\int x^n \,\mathrm{d}x = \frac{x^{n+1}}{n+1}, \qquad n \neq 1$$

$$\int \frac{\mathrm{d}x}{x} = \ln|x|$$

$$\int \frac{f'(x)}{f(x)} \mathrm{d}x = \ln f(x)$$

$$\int (ax+b)^n \,\mathrm{d}x = \frac{(ax+b)^{n+1}}{a(n+1)}, \qquad n \neq 1$$

$$\int \frac{\mathrm{d}x}{ax+b} = \frac{1}{a}\ln(ax+b)$$

$$\int e^{ax}\,\mathrm{d}x = \frac{1}{a}e^{ax}$$

$$\int a^x \,\mathrm{d}x = \frac{a^x}{\ln a}, \quad \text{when } a > 0$$

$$\int \sin x \,\mathrm{d}x = -\cos x$$

$$\int \cos x \, dx = \sin x$$

$$\int \sec^2 x \, dx = \tan x$$

$$\int \tan x \, dx = \ln \sec x$$

$$\int \frac{dx}{\sqrt{(a^2 - x^2)}} = \sin^{-1}\frac{x}{a} \quad \text{when } x^2 < a^2$$

$$\int \frac{dx}{a^2 + x^2} = \frac{1}{a}\tan^{-1}\frac{x}{a}$$

$$\int \frac{dx}{\sqrt{(a^2 + x^2)}} = \sinh^{-1}\frac{x}{a}$$

$$\int \frac{dx}{\sqrt{(x^2 - a^2)}} = \cosh^{-1}\frac{x}{a} \quad \text{when } x > a$$

$$\int \sinh x \, dx = \cosh x$$

$$\int \cosh x \, dx = \sinh x$$

$$\int \tanh x \, dx = \ln \cosh x$$

Examples

1 Find the indefinite integrals with respect to x of

(i) $\dfrac{1}{\sqrt{18 - 2x^2}}$ (ii) $\dfrac{9x^2 + 6}{x^3 + 2x + 1}$ (iii) $\sec^2 3x$

(iv) $\dfrac{1}{5 + 2x^2}$

(i) $\dfrac{dx}{\sqrt{(18 - 2x^2)}}$

$$= \frac{1}{\sqrt{2}}\int \frac{dx}{(9 - x^2)} \qquad \text{standard integral } \int \frac{dx}{\sqrt{(a^2 - x^2)}}$$

$$= \sin^{-1}\frac{x}{a} + c$$

$= \frac{1}{2}\sin^{-1}\dfrac{x}{3} + c$ when c is an arbitrary constant and $x < 3$

(ii) $\displaystyle\int \frac{9x^2+6}{x^3+2x+1}\,dx$

$= 3\displaystyle\int \frac{3x^2+2}{x^3+2x+1}\,dx$ standard integral $\displaystyle\int \frac{f'(x)}{f(x)}\,dx$

$= \ln f(x) + c$

$= 3\ln(x^3+2x+1)+c$, which may be rewritten as

$= \mathbf{3\ln k(x^{+}+2x+1)}$ k an arbitrary constant

(iii) $\displaystyle\int \sec^2 3x\,dx$ standard integral $\displaystyle\int \sec^2 x\,dx$

$= \tan x + c$

$= \dfrac{\tan 3x}{3} + c$

(iv) $\displaystyle\int \frac{dx}{5+2x^2}$

$= \frac{1}{2}\displaystyle\int \frac{dx}{\frac{5}{2}+x^2}$ standard integral $\displaystyle\int \frac{dx}{a^2+x^2}$

$= \dfrac{1}{a}\tan^{-1}\dfrac{x}{a} + c$

$= \frac{1}{2}\tan^{-1}\dfrac{x}{\sqrt{\frac{5}{2}}} + c$

$= \frac{1}{2}\tan^{-1}\sqrt{\frac{5}{2}}x + c$

2 Evaluate (i) $\displaystyle\int_0^{\pi/4} \frac{\cos 2x\,dx}{1+\sin 2x}$ (ii) $\displaystyle\int_{-1}^{0} e^{-2x}\,dx$

(iii) $\displaystyle\int_0^1 (3x-2)^2\,dx$

(i) $\displaystyle\int_0^{\pi/4} \frac{\cos 2x\,dx}{1+\sin 2x}$ standard integral $\displaystyle\int \frac{f'(x)}{f(x)}\,dx$

$= \ln f(x)$

$= \frac{1}{2}[\ln(1+\sin 2x)]_0^{\pi/4}$

$= \frac{1}{2}\ln 2$

(ii) $\int_{-1}^{0} e^{-2x}\,dx$ standard integral $\int e^x\,dx = e^x$

$= \left[\frac{e^{-2x}}{(-2)}\right]_{-1}^{0}$

$= -\frac{1}{2}[e^0 - e^2]$

$= -\frac{1}{2}[1 - e^2]$

$= \frac{1}{2}[e^2 - 1]$

(iii) $\int_0^1 (3x-2)^3\,dx$ standard integral $\int (ax+b)^n\,dx = \frac{(ax+b)n+1}{a(n+1)}$

$= \left[\frac{3(x-2)^4}{3 \times 4}\right]_0^1$

$= \frac{1}{12}[1^4 - (-2)^4] = \frac{1}{12}[1 - 16]$

$= -\frac{15}{12} = -\frac{5}{4}$

Practice questions 16

1 Find the indefinite integrals with respect to x of

(i) $\frac{1}{16+x^2}$ (ii) $\cos 4x$ (iii) e^{7x} (iv) $\tan 5x$

(v) $\frac{3x^2}{x^3+4}$ (vi) $\frac{1}{\sqrt{(50-2x^2)}}$, $|x| < 5$ (vii) $\frac{1}{6+7x^2}$

(viii) $\frac{x^3+x}{x^4+2x^2-5}$ (ix) $\sin 5x$ (x) $(3x+2)^{4/5}$

2 Evaluate

(i) $\int_0^{\pi/3} \frac{\cos 3x}{\sin 3x + 1}\,dx$ (ii) $\int_0^1 e^{3x}\,dx$ (iii) $\int_1^5 (x-1)^{3/2}\,dx$

3 Evaluate the following indefinite integrals

(i) $\int \cosh 2x\,dx$ (ii) $\int \frac{dx}{\sqrt{6+5x^2}}$ (iii) $\int \frac{dx}{\sqrt{x^2-2}}$, $x > \sqrt{2}$

(iv) $\int \tanh 2x\,dx$

Further trigonometric integrals
The following identities are used frequently in finding trigonometric integrals

$$\sin^2 x + \cos^2 x = 1$$

$$\cos 2x = 2\cos^2 x - 1 = 1 - 2\sin x$$

$$2\sin x \cos y = \sin(x + y) + \sin(x - y)$$

$$2\sin x \sin y = \cos(x - y) - \cos(x + y)$$

$$2\cos x \cos y = \cos(x - y) + \cos(x + y)$$

The following examples illustrate the technique used.

Examples

$$\int \cos^2 5x \, dx$$

$$= \frac{1}{2}\int (\cos 10x + 1)\, dx$$

$$= \frac{1}{2}\left[\frac{\sin 10x}{10} + x\right] + c$$

$$= \frac{1}{5}\sin 10x + \frac{1}{2}x + c$$

$$\int \sin 2x \sin 3x \, dx$$

$$= \int \sin 3x \sin 2x \, dx$$

$$= \frac{1}{2}\int (\cos x - \cos 5x)\, dx$$

$$= \frac{1}{2}\left[\sin x - \frac{\sin 5x}{5}\right] + c$$

$$= \frac{1}{2}\sin x - \frac{1}{10}\sin 5x + c$$

$$\int \sin^3 2x \cos^5 2x \, dx$$

$$= \int \sin^2 2x \cos^5 2x \sin 2x \, dx$$

$$= \int (1 - \cos^2 2x)\cos^5 2x \sin 2x \, dx$$

$$= \int (\cos^5 2x \sin 2x - \cos^7 2x \sin 2x)\, dx$$

$$= -\frac{\cos^6 2x}{6 \times 2} + \frac{\cos^8 2x}{8 \times 2} + c$$

$$= -\tfrac{1}{12}\cos^6 2x + \tfrac{1}{16}\cos^8 2x + c$$

Practice questions 17

Evaluate the following

1 $\int \sin^2 3x \, dx$

2 $\int \cos^5 x \, dx$

3 $\int \sin 5x \cos 2x \, dx$

4 $\int \cos 3x \cos 2x \, dx$

5 $\int \cos^5 x \sin^2 x \, dx$

Integration when the integrand is a rational fraction

When the denominator of the integrand factorises, the integrand is expressed in partial fractions before integrating.

Example

Find the following integral

$$\int \frac{7x - 2}{x^2 + x - 2} dx$$

Let $\dfrac{7x - 2}{x^2 + x - 2} \equiv \dfrac{A}{x - 1} + \dfrac{B}{x + 2}$

$\therefore \; 7x - 2 \equiv A(x + 2) + B(x - 1)$

when $x = 1$ $\quad 5 = 3A$

$$A = \frac{5}{3}$$

when $x = -2$ $\quad -16 = -3B$

$$B = \frac{16}{3}$$

$$\int \frac{7x - 2}{x^2 + x - 2} dx = \frac{5}{3}\int \frac{dx}{x - 1} + \frac{16}{3}\int \frac{dx}{x + 2}$$

$$= \frac{5}{3}\ln(x - 1) + \frac{16}{3}\ln(x + 2) + c$$

When the denominator of the integrand does not factorise and is quadratic we change the integral into the form $f'x/f(x) + I_1$, where I_1 is reducible to a standard form.

Example

Find
$$\int \frac{x+7}{x^2-3x+4}\,dx$$

$$\frac{d}{dx}(x^2-3x+1) = 2x-3$$

$$\frac{x+7}{x^2-3x+4} \equiv \frac{1}{2}\cdot\frac{2x+14}{x^2-3x+4} \equiv \frac{1}{2}\cdot\frac{2x-3+3+14}{x^2-3x+14}$$

$$\equiv \frac{1}{2}\cdot\frac{2x-3}{x^2-3x+4} + \frac{1}{2}\cdot\frac{17}{x^2-3x+4}$$

$$\int \frac{x+7}{x^2-3x+4}\,dx = \frac{1}{2}\int \frac{2x-3}{x^2-3x+4}\,dx + \frac{17}{2}\int \frac{dx}{x^2-3x+4}$$

$$= \frac{1}{2}\ln(x^2-3x+4) + \frac{17}{2}\int \frac{dx}{x^2-3x+\left(\frac{3}{2}\right)^2-\left(\frac{3}{2}\right)^2+4}$$

$$= \frac{1}{2}\ln(x^2-3x+4) + \frac{17}{2}\int \frac{dx}{\left(x-\frac{3}{2}\right)^2+\frac{7}{4}}$$

$$= \frac{1}{2}\ln(x^2-3x+4) + \frac{17}{2}\,\frac{2}{\sqrt{7}}\tan^{-1}\left(\frac{x-\frac{3}{2}}{\frac{\sqrt{7}}{2}}\right) + c$$

$$= \frac{1}{2}\ln(x^2-3x+4) + \frac{17}{\sqrt{7}}\tan^{-1}\frac{2x-3}{\sqrt{7}} + c$$

Practice questions 18

Find

1 $\int \frac{dx}{x^2-2x}$ 2 $\int \frac{x+1}{x^2+1}\,dx$ 3 $\int \frac{3x+4}{x^2-7x+12}$

5 $\int \frac{2x+5}{x^2+3x+4}\,dx$ 6 $\int \frac{x-4}{x^2+x-8}\,dx$ 7 $\int \frac{x^3}{x^2+x+1}\,dx$

Integration by change of variable

Some integrals can be evaluated more easily by the substitution of a new variable u in place of the original variable x. Suppose that $I = \int f(x)\,dx$ and we wish to change the variable from x to u by means of the substitution $x = h(u)$ and that this relationship changes $f(x)$ into $F(u)$.

Since
$$I = \int f(x)\,dx$$
$$\frac{dI}{dx} = f(x)$$
$$= F(u)$$
$$\frac{dI}{du} = \frac{dI}{dx}\frac{dx}{du}$$
$$= F(u)\frac{dx}{du}$$
$$\therefore I = \int F(u)\frac{dx}{du}\,du$$

i.e. $f(x)$ is replaced by $F(u)$ and dx by $dx/du\,du$.

Examples

Find $I = \int x\sqrt{1-x}\,dx, \quad x < 1$

Let $u = 1 - x$

differentiating w.r. to u,

$I = -\dfrac{dx}{du}$ i.e. $\dfrac{dx}{du} = -1$

$$I = \int (1-u)\sqrt{u}(-1)\,du$$
$$= \int (u^{3/2} - u^{1/2})\,du$$
$$= \frac{2}{5}u^{5/2} - \frac{2}{3}u^{3/2} + C$$
$$= \frac{2}{5}(1-x)^{5/2} - \frac{2}{3}(1-x)^{3/2} + C$$

2 Evaluate $\displaystyle\int_2^{2\sqrt{3}} \sqrt{16 - x\,dx}$

let $x = 4\sin\theta$

then $dx/d\theta = 4\cos\theta$

$$16 - x^2 = 16 - 16\sin^2\theta = 16\cos^2\theta$$
$$\sqrt{16 - x^2} = 4\cos\theta$$

also when $x = 2$, $\sin\theta = \dfrac{1}{2}$

$$\theta = \frac{\pi}{6}$$

when $x = 2\sqrt{3}$, $\sin\theta = \dfrac{\sqrt{3}}{2}$

$$\theta = \frac{\pi}{3}$$

$$I = \int_{\pi/6}^{\pi/3} 4\cos\theta \, 4\cos\theta \, d\theta$$

$$= 16\int_{\pi/6}^{\pi/3} \cos^2\theta \, d\theta$$

$$= 8\int_{\pi/6}^{\pi/3} (1 + \cos 2\theta)\, d\theta$$

$$= 8\left[\theta + \frac{\sin 2\theta}{2}\right]_{\pi/6}^{\pi/3}$$

$$= 8\left[\frac{\pi}{3} + \frac{\sqrt{3}}{4}\right] - 8\left[\frac{\pi}{6} + \frac{\sqrt{3}}{4}\right]$$

$$= 8 \, . \, \frac{\pi}{6}$$

$$= \frac{4\pi}{3}$$

Practice questions 19

1 $\displaystyle\int x\sqrt{1+x}\;\; dx, \quad x > -1$ 2 $\displaystyle\int x^2\sqrt{1-x}\, dx, \quad x > 1$

3 $\displaystyle\int_{\theta}^{3/\sqrt{2}} \sqrt{9 - x^2}\, dx,$ 4 $\displaystyle\int_0^5 x^2\sqrt{25 - x^2}\, dx$

5 $\displaystyle\int \frac{dx}{x^2\sqrt{9 - x^2}}; \quad |x| < 3$

The 't' substitution

This substitution is used to change a trigonometrical integrand into an algebraic integrand.

We let $t = \tan\dfrac{x}{2}$ then

$$\sin x = \frac{2t}{1+t^2}, \quad \cos x = \frac{1-t^2}{1+t^2} \quad \text{and} \quad \tan x = \frac{2t}{1+t^2}$$

$$\mathrm{d}x = \frac{2\,\mathrm{d}t}{1+t^2}$$

Example

Find $I = \displaystyle\int_0^{\pi/3} \frac{\mathrm{d}x}{3+\cos x}$

Let $t = \tan\dfrac{x}{2}$

then $\cos x = \dfrac{1-t^2}{1+t^2}$ and $\mathrm{d}x = \dfrac{2\,\mathrm{d}t}{1+t^2}$ when $x = 0$, $t = 0$,

when $x = \dfrac{\pi}{3}$, $t = \dfrac{1}{\sqrt{3}}$

$$I = \int_0^{1/\sqrt{3}} \frac{\dfrac{2\,\mathrm{d}t}{1+t^2}}{3+\left(\dfrac{1-t^2}{1+t^2}\right)}$$

$$= \int_0^{1/\sqrt{3}} \frac{2\,\mathrm{d}t}{3(1+t^2)+(1-t^2)}$$

$$= \int_0^{1/\sqrt{3}} \frac{2\,\mathrm{d}t}{4+2t^2}$$

$$= \int_0^{1/\sqrt{3}} \frac{\mathrm{d}t}{2+t^2}$$

$$= \left[\frac{1}{\sqrt{2}}\tan^{-1}\frac{t}{\sqrt{2}}\right]_0^{1/\sqrt{3}}$$

$$= \frac{1}{\sqrt{2}}\tan^{-1}\frac{1}{\sqrt{6}}$$

$= 0{\cdot}274$ **to 3 decimal places**

Practice questions 20

1 Find the value of $\displaystyle\int_0^{\pi/2} \frac{6\,\mathrm{d}\theta}{4\sin\theta + 3\cos\theta + 5}$

2 Evaluate $\displaystyle\int_0^{2\pi/3} \frac{dx}{2 + \cos x}$

3 Evaluate $\displaystyle\int_0^{\pi/2} \frac{dx}{1 + \sin x}$

Integration by parts

Integration by parts is the method used to integrate the product of two functions of the form $u \cdot \dfrac{dv}{dx}$.

The product rule for the differentiation of two functions u and v states that:

$$\frac{d}{dx}(uv) = v\frac{du}{dx} + u\frac{dv}{dx}$$

Integrating with respect to x,

$$\int \frac{d}{dx}(uv)\,dx = \int v\frac{du}{dx}\,dx + \int u\frac{dv}{dx}\,dx$$

$$uv = \int v\frac{du}{dx}\,dx + \int u\frac{dv}{dx}\,dx$$

$$\therefore \quad \int \mathbf{u}\frac{\mathbf{dv}}{\mathbf{dx}}\,\mathbf{dx} = \mathbf{uv} - \int \mathbf{v}\frac{\mathbf{du}}{\mathbf{dx}}\,\mathbf{dx}$$

or in differential form

$$\int \mathbf{u\,dv} = \mathbf{uv} - \int \mathbf{v\,du}$$

The formula is also used when the integrand contains an inverse trigonometrical or hyperbolic function or a logarithmic function.

Examples

Find

(i) $\int x^3 \ln x\,dx$ (ii) $\int x \sin x\,dx$ (iii) $\int \sin^{-1} 2x\,dx$

(iv) $\int x^2 e^{3x}\,dx$ (v) $\int e^{2x} \cos 3x\,dx$ (vi) $\int \sqrt{4 + x^2}\,dx$

(i) $I = \int x^3 \ln x\,dx$

Since $\ln x$ cannot be integrated immediately choose x^3 to be equal to dv/dx, then

$$\frac{dv}{dx} = x^3$$

$$v = \frac{x^4}{4}$$

N.B. We choose the simplest v, i.e. that with no arbitrary constant.

$$I = \int \ln x \, . \, x^3 \, dx$$

$$= \int \ln x \, . \, \frac{d}{dx}\left(\frac{x^4}{4}\right) dx$$

$$= \ln x \, . \, \frac{x^4}{4} - \int \frac{x^4}{4} \, . \, \frac{1}{x} \, dx$$

$$= \frac{x^4}{4} \ln x - \frac{1}{4} \int x^3 \, dx$$

$$= \frac{x^4}{4} \ln x - \frac{x^4}{16} + c$$

$$= \frac{x^4}{16}(4 \ln x - 1) + c$$

(ii) $I = \int x \sin x \, dx$

here both functions can be integrated and differentiated immediately. Since $\frac{d}{dx}(x) = 1$ we choose x to be u then

$$\sin x = \frac{dv}{dx}$$

$$-\cos x = v$$

$$I = \int x \sin x \, dx$$

$$= \int x \frac{d}{dx}(-\cos x) \, dx$$

$$= x(-\cos x) - \int (-\cos x) \, . \, 1 \, dx$$

$$= -x \cos x + \int \cos x \, dx$$

$$= -x \cos x + \sin x + c$$

(iii) $I = \int \sin^{-1} 2x \, dx$

$$= \int \sin^{-1} 2x \, . \, 1 \, dx$$

choose $\dfrac{dv}{dx} = 1$ then $v = x$

$$I = \int \sin^{-1} 2x \,.\, \frac{d}{dx}(x)\,dx$$

$$= (\sin^{-1} 2x)x - \int x \,.\, \frac{1}{\sqrt{1-4x^2}} \,.\, 2\,dx$$

$$= x \sin^{-1} 2x - 2\int \frac{x}{\sqrt{1-4x^2}}\,dx$$

$$= x \sin^{-1} 2x - 2\int (1-4x^2)^{-1/2} x\,dx$$

$$= x \sin^{-1} 2x + \frac{1}{4}\int (1-4x^2)^{-1/2}(-8x)\,dx$$

$$= x \sin^{-1} 2x + \frac{1}{2}(1-4x^2)^{1/2} + c$$

(iv) $I = x^2 e^{3x}\,dx$

Note that $\dfrac{d^2}{dx^2}(x^2) = 2$ so choose $e^{3x} = \dfrac{dv}{dx}$

then $v = \dfrac{1}{3}e^{3x}$

$$I = \int x^2 \frac{d}{dx}\left(\frac{1}{3}e^{3x}\right)dx$$

$$I = x^2\left(\frac{1}{3}e^{3x}\right) - \frac{2}{3}\int xe^{3x}\,dx \qquad (1)$$

Let $I_1 = \displaystyle\int xe^{3x}\,dx$

again choose $e^{3x} = \dfrac{dv}{dx}$

then $I_1 = \displaystyle\int x \frac{d}{dx}\left(\frac{1}{3}e^{3x}\right)dx$

$$= x\left(\frac{1}{3}e^{3x}\right) - \int \frac{1}{3}e^{3x} \,.\, 1\,dx$$

$$I_1 = \frac{1}{3}xe^{3x} - \frac{1}{9}e^{3x} + k, \ k \text{ an arbitrary constant} \qquad (2)$$

Substituting (2) in (1)

$$I = \frac{1}{3}x^2e^{3x} - \frac{2}{3}\left\{\frac{1}{3}xe^{3x} - \frac{1}{9}e^{3x} + k\right\}$$

$$= \frac{1}{3}x^2e^{2x} - \frac{2}{9}xe^{3x} - \frac{2}{27}e^{3x} + c, \; c \text{ an arbitrary constant}$$

(v) $I = \int e^{2x} \cos 3x \, dx$

Here both functions can be integrated immediately and also can be differentiated immediately. We choose $e^{2x} = \frac{dv}{dx}$ [the reader is advised to evaluate this integral by choosing $\frac{dv}{dx} = \cos 3x$]

$$\frac{dv}{dx} = e^{2x} \quad v = \frac{1}{2}e^{2x}$$

$$I = \int \cos 3x \frac{d}{dx}\left(\frac{1}{2}e^{2x}\right) dx$$

$$= \cos 3x\left(\frac{1}{2}e^{2x}\right) - \int \frac{1}{2}e^{2x}(-3 \sin 3x)\, dx$$

$$\therefore \; I = \frac{1}{2}e^{2x} \cos 3x + \frac{3}{2}\int e^{2x} \sin 3x \, dx \tag{1}$$

Let $I_1 = \int e^{2x} \sin 3x \, dx$

Since we chose $e^{2x} = dv/dx$ in the first application of the formula we must do so again. Thus

$$I_1 = \int \sin 3x \, . \frac{d}{dx}\left(\frac{1}{2}e^{2x}\right) dx$$

$$= \sin 3x\left(\frac{1}{2}e^{2x}\right) - \int \frac{1}{2}e^{2x}(3 \cos 3x)\, dx$$

$$= \frac{1}{2}e^{2x} \sin 3x - \frac{3}{2}\int e^{2x} \cos 3x \, dx$$

$$\therefore \; I_1 = \frac{1}{2}e^{2x} \sin 3x - \frac{3}{2}I \tag{2}$$

substituting (2) in (1)

$$I = \frac{1}{2}e^{2x} \cos 3x + \frac{3}{2}\left\{\frac{1}{2}e^{2x} \sin 3x - \frac{3}{2}I\right\}$$

$$I = \frac{1}{2}e^{2x} \cos 3x + \frac{3}{4}e^{2x} \sin 2x - \frac{9}{4}I$$

$$\frac{13}{14}I = \frac{1}{4}e^{2x}(2\cos 3x + 3\sin 2x)$$

$$\mathbf{I = \frac{1}{13}e^{2x}(2\cos 3x + 3\sin 2x) + c,}$$

introducing the arbitrary constant to give the most general form of I.

(vi) $I = \int_0^1 \sqrt{4 + x^2}\,dx$

choose $\frac{dv}{dx} = 1$

then $\quad I = \int_0^1 \sqrt{4+x^2}\frac{d}{dx}(x)\,dx$

$$= [x\sqrt{4+x^2}]_0^1 - \int_0^1 x\,.\frac{1}{2}(4+x^2)^{1/2}\,.\,2x\,dx$$

$$= [\sqrt{5}] - [0] - \int_0^1 \frac{x^2}{\sqrt{4+x^2}}dx$$

$$= \sqrt{5} - \int_0^1 \frac{4+x^2-2}{\sqrt{4+x^2}}dx$$

$$= \sqrt{5} - I + 4\int_0^1 \frac{dx}{\sqrt{4+x^2}}$$

$$\therefore\ I = \sqrt{5} - I + 4\left[\sinh^{-1}\frac{x}{2}\right]_0^1$$

$$2I = \sqrt{5} + 4[\sinh^{-1}\frac{1}{2} - 0]$$

$$I = \frac{1}{2}\sqrt{5} + 2\sinh^{-1}\frac{1}{2}$$

$$= \frac{1}{2}\sqrt{5} + 2\ln\left(\frac{1}{2} + \frac{\sqrt{5}}{2}\right)$$

Practice questions 21

Find the values of

1 $\int x \ln x\,dx$ $\qquad$ 2 $\int x\cos x\,dx$ $\qquad$ 3 $\int x^2 \sin x\,dx$

4 $\int e^{-x} \cos 3x \, dx$ 5 $\int \sin^{-1} x \, dx$ 6 $\int \sec^3 x \, dx$

7 $\int_0^1 x^2 e^{-x} \, dx$

Reduction formulae

A reduction formula expresses an integral I_n in terms of a similar integral where n is reduced by 1 or more. Repeated applications of the formula enables I_n to be evaluated for particular values of n. The reduction formula is obtained by using the method of integrating by parts on I_n.

Examples

1 Find a reduction formula for $\int \cos^n x \, dx$, where n is a positive integer.

$$I_n = \int \cos^n x \, dx$$

$$= \int \cos^{n-1} x \,.\, \cos x \, dx$$

$$= \int \cos^{n-1} x \frac{d}{dx}(\sin x) \, dx$$

$$= \sin x \,.\, \cos^{n-1} x - \int \sin x \,.\, (n-1)\cos^{n-2} x(-\sin x) \, dx$$

$$= \sin x \cos^{n-1} x + (n-1) \int \cos^{n-2} x \sin^2 x \, dx$$

$$= \sin x \cos^{n-1} x + (n-1) \int \cos^{n-2} x(1 - \cos^2 x) \, dx$$

$$= \sin x \cos^{n-1} x + (n-1) \int (\cos^{n-2} x - \cos^n x) \, dx$$

$$\therefore \; I_n = \sin x \cos^{n-1} x + (n-1)I_{n-2} - (n-1)I_n$$

$$nI_n = \sin x \cos^{n-1} x + (n-1)I_{n-2}$$

$$I_n = \frac{1}{n} \sin x \cos^{n-1} x + \frac{(n-1)}{n} I_{n-2}$$

2 Obtain a reduction formula for

$$\int_{a(\pi/2)}^{b(\pi/2)} \cos^m x \sin^n x \, dx$$

where a and b are integers and $b > a$, and m and n are positive integers.

$$I_{m,n} = \int_{a(\pi/2)}^{b(\pi/2)} \cos^m x \sin^n x \, dx$$

$$= \int_{a(\pi/2)}^{b(\pi/2)} \cos^{m-1} x \sin^n x \cos x \, dx$$

Let $\dfrac{dv}{dx} = \sin^n x \cos x$

then $v = \dfrac{1}{n+1} \sin^{n+1} x$

$$I_{m,n} = \int_{a(\pi/2)}^{b(\pi/2)} \cos^{m-1} x \frac{d}{dx}\left(\frac{1}{n+1}\sin^{n+1} x\right) dx$$

$$= \left[\frac{1}{n+1}\sin^{n+1} x \cos^{m-1} x\right]_{a(\pi/2)}^{b(\pi/2)} - \int_{a(\pi/2)}^{b(\pi/2)} \frac{1}{n+1}$$
$$\times \sin^{n+1} x(m-1)\cos^{m-2} x(-\sin x)\, dx$$

$$= 0 + \frac{m-1}{n+1}\int_{a(\pi/2)}^{b(\pi/2)} \cos^{m-2} x \sin^{n+2} x \, dx$$

The power of $\cos x$ has been reduced but the power of $\sin x$ has been increased, but rewriting $\sin^{n+2} x$ as $\sin^n x \sin^2 x$ we obtain

$$I_{m,n} = \frac{m-1}{n+1}\int_{a(\pi/2)}^{b(\pi/2)} \cos^{m-2} x \sin^n x \sin^2 x \, dx$$

$$= \frac{m-1}{n+1}\int_{a(\pi/2)}^{b(\pi/2)} \cos^{m-2} x \sin^n x (1 - \cos^2 x)\, dx$$

$$= \frac{m-1}{n+1}\int_{a(\pi/2)}^{b(\pi/2)} (\cos^{m-2} x \sin^n x - \cos^m x \sin^n x)\, dx$$

$$\therefore\ I_{m,n} = \frac{m-1}{n+1}\{I_{m-2,n} - I_{m,n}\}$$

$$(n+1)I_{m,n} = (m-1)\{I_{m-2,n} - I_{m,n}\}$$

$$(n+1)I_{m,n} + (m-1)I_{m,n} = (m-1)I_{m-2,n}$$

$$(m+n)I_{m,n} = (m-1)I_{m-2,n}$$

$$I_{m,n} = \frac{m-1}{m+n} I_{m-2,n}$$

3 Use the reduction formula of Question 2 to evaluate

$$\int_0^{\pi/2} \cos^5 x \sin^3 x \, dx$$

In this case $a = 0$ and $b = 1$

$$I_{m,n} = \frac{m-1}{m+n} I_{m-2,n}$$

$$I_{5,3} = \frac{4}{8} I_{3,3}$$

$$I_{3,3} = \frac{2}{6} I_{1,3}$$

$$\therefore\ I_{5,3} = \frac{4}{8} \cdot \frac{2}{6} \int_0^{\pi/2} \cos x \sin^3 x \, dx$$

$$= \frac{1}{6} \left[\frac{\sin^4 x}{4} \right]_0^{\pi/2}$$

$$= \frac{1}{24}$$

Practice questions 22

1 Find a reduction formula for $\int_0^{\pi/2} \sin^n x \, dx$ where n is a positive integer greater than 1. Hence find the value of

(i) $\int_0^{\pi/2} \sin^5 x \, dx$ (ii) $\int_0^{\pi/2} \sin^8 x \, dx$

2 Find a reduction formula for $\int \sec^n x \, dx$ where n is a positive integer greater than 1. Hence evaluate $\int \sec^4 x \, dx$.

3 Find a reduction formula for $\int x^n e^{ax} \, dx$. Hence find the value of $\int_0^1 x^4 e^{-x} \, dx$.

Applications of integration

1 *Areas*

If $f(x)$ is positive and is integrable on $a \leq x \leq b$, then the area, A, of the region bounded by the curve $y = f(x)$, the ordinates $x = a$ and $x = b$, and the x-axis is given by

$$A = \int_a^b f(x) \, dx$$

If $f(x)$ is negative on $a \leq x \leq b$, the corresponding region is below the x-axis. The value of the integral is now negative and the area of the region, A_1, is given by the numerical value of the integral i.e.

$$A_1 = \left| \int_a^b f(x) \, dx \right|$$

When finding the area of a region bounded by the curve $y = F(x)$, the ordinates $x = p$, and $x = q$, and the x-axis, determine firstly whether the curve cuts the x-axis anywhere between $x = p$, and $x = q$. If it does then determine the values of the areas above and below the x-axis separately, adding together the numerical values obtained to obtain the total area.

Examples

1 Find the area enclosed by the curve $y = x^3 - 5x^2 + 4x$ and the x-axis between $x = 0$ and $x = 4$.

$$y = x^3 - 5x^2 + 4x$$
$$= x(x-1)(x-4)$$

curve cuts the x-axis at $x = 0$, 1 and 4.

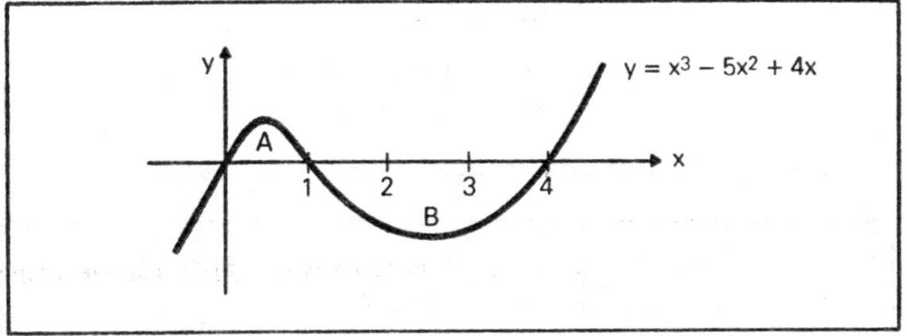

$$\text{Area } A = \int_0^1 (x^3 - 5x^2 + 4x)\,dx$$
$$= \left[\frac{x^4}{4} - \frac{5x^3}{3} + 2x^2\right]_0^1$$
$$= \frac{7}{12}\text{ units}^2$$

$$\text{Area } B = \left|\int_1^4 (x^3 - 5x^2 + 4x)\,dx\right|$$
$$= \left|\left[\frac{x^4}{4} - \frac{5x^3}{3} + 2x^2\right]_1^4\right|$$
$$= \left|64 - \frac{5}{3}.64 + 32 - \frac{7}{12}\right|$$
$$= \left|-\frac{45}{4}\right|$$
$$= \frac{45}{4}\text{ units}^2$$

Total area enclosed is $\frac{7}{12} + \frac{45}{4}$ units^2

$$= \frac{142}{12} \text{ units}^2$$

2 Find the area enclosed between the curve $y = x(x-1)(x-2)$ and the straight line $3x - y - 3 = 0$.
The curve $y = x(x-1)(x-2)$ cuts the x-axis at $x = 0, 1, 2$ and y is positive when $0 < x < 1$ and when $x > 2$. The straight line $3x - y - 3 = 0$ cuts the curve when

$$x(x-1)(x-2) = 3x - 3$$
$$(x-1)[x(x-2) - 3] = 0$$
$$(x-1)(x^2 - 2x - 3) = 0$$
$$(x-1)(x+1)(x-3) = 0$$
$$x = -1, 1, 3$$

i.e. straight line and curve intersect at $(-1, -6)$ $(1, 0)$ and $(3, 6)$.

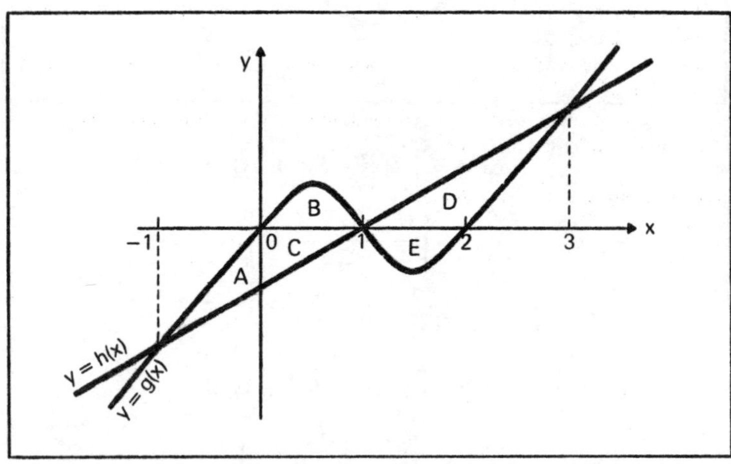

The area enclosed between the curve and the straight line is the numerical sum of the areas A, B, C, D and E in the diagram. Let $g(x) = x(x-1)(x-2)$ and $h(x) = 3x - 3$, then areas enclosed =

$$\left|\int_{-1}^{0} (h(x) - g(x))\,dx\right| + \int_{0}^{1} g(x)\,dx + \left|\int_{0}^{1} h(x)\,dx\right|$$
$$+ \int_{1}^{3} h(x)\,dx - \int_{2}^{3} g(x)\,dx + \left|\int_{1}^{2} g(x)\,dx\right|$$

$$= \int_{-1}^{1} \{g(x) - h(x)\}\,dx + \int_{1}^{3} \{h(x) - g(x)\}\,dx$$

$$= \int_{-1}^{1} \{x(x-1)(x-2) - (3x-3)\}\,dx$$

$$+ \int_{1}^{3} \{(3x-3) - x(x-1)(x-2)\}\,dx$$

$$= \int_{-1}^{1} (x^3 - 3x^2 - x + 3)\,dx + \int_{1}^{3} (-x^3 + 3x^2 + x - 3)\,dx$$

$$= \left[\frac{x^4}{4} - x^3 - \frac{x^2}{2} + 3x\right]_{-1}^{1} + \left[-\frac{x^4}{4} + x^3 + \frac{x^2}{2} - 3x\right]_{1}^{3}$$

$$= 1\tfrac{3}{4} + 2\tfrac{1}{4} + 2\tfrac{1}{4} + 1\tfrac{3}{4} = 8$$

Area is 8 units^2.

Practice questions 23

1 Find the region bounded by the curve $y = \cos x$ and the x-axis between $x = -\pi/2$ and $x = 3\pi/2$.

2 Find the area enclosed between the curve $y = (x + 1)(x + 2)(x + 4)$ and the x-axis.

3 Find the area enclosed between the line $y = x + 3$ and the curve $y = 1 + \cos x$ and the ordinates $x = 0$ and $x = \pi/2$.

Area in terms of polar coordinates

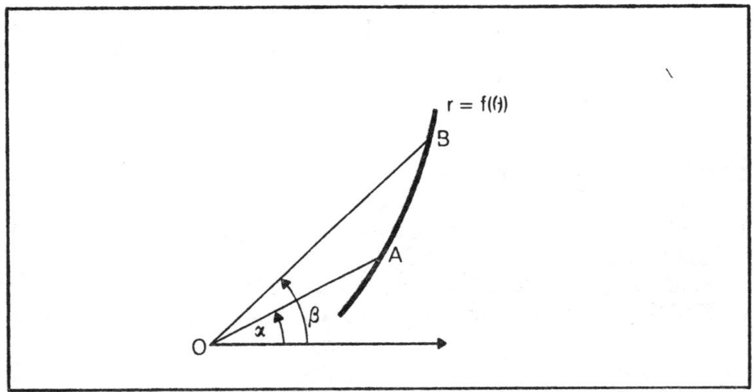

The area of the sector OAB bounded by the continuous curve $r = f(\theta)$ and the lines $\theta = x\alpha$ and $\theta = \beta$ where $\alpha \le \theta \le \beta$ and $\beta < \alpha + 2\pi$ is given by

$$\int_\alpha^\beta \frac{1}{2}r^2\, d\theta$$

Example
Find the total area enclosed by the 4-leaved rose whose equation is $r = \alpha \cos 2\theta$.

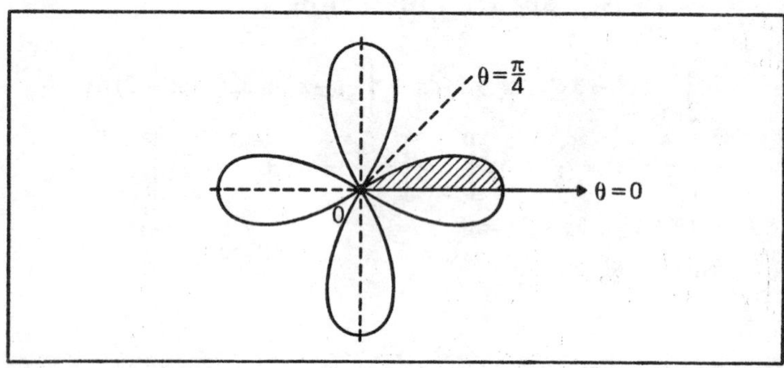

The required area is eight times the area shaded.

$$\text{Total area enclosed} = 8\int_0^{\pi/4} \frac{1}{2}a^2 \cos^2 2\theta\, d\theta$$

$$= 4a^2 \int_0^{\pi/4} \frac{1 + \cos 4\theta}{2}\, d\theta$$

$$= 2a^2 \left[\theta + \frac{1}{4}\sin 4\theta\right]_0^{\pi/4}$$

$$= \frac{1}{2}a^2\pi \text{ units}^2.$$

Practice questions 24

1 Find the total area enclosed by the curve $r = 4\sin^2\theta$.
2 Find the area bounded by the curve $r = 2(1 - \cos\theta)$ and the lines $\theta = 0$ and $\theta = \pi/2$.
3 Find the area of the lemniscate $r^2 = 25\cos 2\theta$.
4 Find the area of the region common to the circle $r = 2$ and the cardioid $r = 2(1 - \cos\theta)$.

2 *Volume of a solid of revolution*
The volume of the solid generated by the revolution about the x-axis of the part of the continuous curve $y = f(x)$ which lies between $x = a$ and $x = b$ where $a < x < b$ is given by

$$\int_a^b \pi[f(x)]^2\,dx$$

Example
Find the volume of the solid formed when the area enclosed between the curve $y = \sin x$ and the lines $x = 0$ and $x = \pi/2$ is rotated through one revolution about the x-axis.

$$\text{Volume generated} = \int_0^{\pi/2} \pi \sin^2 x\,dx$$

$$= \int_0^{\pi/2} \frac{1}{2}\pi(1 - \cos 2x)\,dx$$

$$= \frac{1}{2}\pi\left[x - \frac{1}{2}\sin 2x\right]_0^{\pi/2}$$

$$= \frac{1}{2}\pi\left[\frac{\pi}{2}\right]$$

$$= \frac{\pi^2}{4}\ \text{units}^3$$

Practice questions 25

1 Find the formula for the volume of a sphere by considering the rotation of the curve $y = +\sqrt{r^2 - x^2}$ about the x-axis.
2 Find the volume of the solid of revolution formed by rotating the curve $y = +\sqrt{x/(2 - x)}$ between $x = 0$ and $x = 1$ about the x-axis.

3 *Length of arc of a curve*
The length of the arc of a continuous curve $y = f(x)$ between the points on the curve $(a, f(a))$ and $(b, f(b))$ is given by

$$\int_a^b \sqrt{1 + \left(\frac{dy}{dx}\right)^2}\,dx$$

When the curve is given by parametric equations $x = f(t)$, $y = g(t)$ the length of arc from the point where $t = t_1$ to the point where $t = t_2$ and $t_2 > t_1$ is given by

$$\int_{t_1}^{t_2} \sqrt{\left(\frac{dx}{dt}\right)^2 + \left(\frac{dy}{dt}\right)^2}\,dt$$

Examples
1 Find the length of the arc of the curve $y = x^{3/2}$ from $x = 0$ to $x = 5$.

$$y = x^{3/2}$$

$$\frac{dy}{dx} = \frac{3}{2}x^{1/2}$$

$$1 + \left(\frac{dy}{dx}\right)^2 = 1 + \frac{9}{4}x - \frac{1}{4}(4 + 9x)$$

$$\text{Length of arc} = \int_0^5 \sqrt{1 + \left(\frac{dy}{dx}\right)^2}\, dx$$

$$= \int_0^5 \frac{1}{2}\sqrt{4 + 9x}\, dx$$

$$= \frac{1}{2}\left[(4 + 9x)^{3/2} \cdot \frac{2}{3} \cdot \frac{1}{9}\right]_0^5$$

$$= \frac{1}{27}[343 - 8]$$

$$= \frac{335}{27} \text{ units.}$$

2 Find the length of the arc of the curve given by the parametric equations $x = e^t \sin t$, $y = e^t \cos t$ from $t = 0$ to $t = 1$.

$$x = e^t \sin t, \qquad y = e^t \cos t$$

$$\frac{dx}{dt} = e^t \sin t + e^t \cos t, \qquad \frac{dy}{dt} = -e^t \sin t + e^t \cos t$$

$$\frac{dx}{dt} = e^t(\sin t + \cos t), \qquad \frac{dy}{dt} = e^t(\cos t - \sin t)$$

$$\left(\frac{dx}{dt}\right)^2 + \left(\frac{dy}{dt}\right)^2 = e^{2t}(\sin t + \cos t)^2 + e^{2t}(\cos t - \sin t)^2$$

$$= 2e^{2t}$$

$$\text{Length of arc} = \int_0^1 \sqrt{\left(\frac{dx}{dt}\right)^2 + \left(\frac{dy}{dt}\right)^2}\, dt$$

$$= \int_0^1 \sqrt{2}\, e^t\, dt$$

$$= \sqrt{2}\,[e^t]_0^1$$

$$= \sqrt{2}[e - 1] \text{ units.}$$

Practice questions 26

1 Find the length of the arc of the curve $y = 1/2\,(e^x + e^{-x})$ between the points where $x = -1$ and $x = 1$.
2 Find the length of arc of the curve $y = \ln x - x^2/8$ between the points where $x = 1$ and $x = 2$.
3 Find the length of the curve $x = \cos^3 t, y = \sin^3 t$ from $\theta = 0$ to $\theta = \pi/2$.
4 Find the length of one arch of the cycloid $x = 3\,(\theta - \sin\theta), y = 3\,(1 = \cos\theta)$.
5 Find the length of the catenary $y = (1/2)\cosh 2x$ from $x = 0$ to $x = a$.

Length of arc in polar coordinates
The length of arc of the curve $r = f(\theta)$ from $r = \alpha$ to $r = \beta$ and $\alpha < \beta$ is given by

$$\int_\alpha^\beta \sqrt{r^2 + \left(\frac{dr}{d\theta}\right)^2}\, d\theta$$

Example
Find the total length of the perimeter of the cardioid $r = 6\,(1 + \cos\theta)$

$$r = 6(1 + \cos\theta)$$

$$\frac{dr}{d\theta} = -6\sin\theta$$

$$r^2 + \left(\frac{dr}{d\theta}\right)^2 = 36(1 + \cos\theta)^2 + 36\sin^2\theta$$

$$= 36(2 + 2\cos\theta)$$

$$= 72\left(2\cos^2\frac{\theta}{2}\right)$$

$$= 144\cos^2\frac{\theta}{2}$$

Since the curve is symmetrical above the initial line the total length of arc will be twice that above the initial line, which lies between θ and 0 and $\theta = \pi$

$$\therefore\ s = 2\int_0^\pi \sqrt{r^2 + \left(\frac{dr}{d\theta}\right)^2}\, d\theta$$

$$= 2\int_0^\pi 12\cos\frac{\theta}{2}\, d\theta$$

$$= 24\left[2\sin\frac{\theta}{2}\right]_0^\pi$$

$$= \mathbf{48\ units.}$$

Practice questions 27

1 Find the length of arc of the equiangular spiral $r = ae^{a\theta}$ from $\theta = 0$ to $\theta = \alpha$.

2 Find the length of the curve $r = \sin^3 t/3$.

4 *Area of the surface of a solid of revolution*

The area of the surface of revolution generated when an arc of a curve $y = f(x)$ is rotated through a revolution about the x-axis is given by

$A = \int_{x_1}^{x_2} 2\pi y \, ds$ where x_1 and x_2 are the x-coordinates of the extremities of the arc.

$$\therefore \; A = 2\pi \int_{x_1}^{x_2} y\sqrt{1 + \left(\frac{dy}{dx}\right)^2}\, dx$$

If the curve is given in terms of a parameter t then

$$A = 2\pi \int_{t_1}^{t_2} y\sqrt{\left(\frac{dx}{dt}\right)^2 + \left(\frac{dy}{dt}\right)^2}\, dt$$

Example

Find the surface of the paraboloid obtained by rotating the arc of the parabola $y^2 = 4x$ between $x = 0$ and $x = 3$ about the x-axis. The surface area is that generated by rotating the arc $y = +2x^{1/2}$ through 1 revolution about Ox.

$$y = 2x^{1/2}$$

$$\frac{dy}{dx} = x^{-1/2}$$

$$1 + \left(\frac{dy}{dx}\right)^2 = 1 + \frac{1}{x}$$

$$\text{Area required} = 2\pi \int_0^3 y\sqrt{1 + \left(\frac{dy}{dx}\right)^2}\, dx$$

$$= 2\pi \int_0^3 2x^{1/2}\left(1 + \frac{1}{x}\right)^{1/2} dx$$

$$= 4\pi \int_0^3 (x + 1)^{1/2}\, dx$$

$$= 4\pi \left[\frac{2}{3}(x + 1)^{3/2}\right]_0^3$$

$$= 4\pi \left[\frac{16}{3} - \frac{2}{3}\right] = \frac{56\pi}{3} \text{ units}^2.$$

Practice questions 28

1 Find the area of that part of the surface of a sphere radius r bounded by two parallel planes distance $\frac{1}{3}r$ and $\frac{1}{2}r$ from the centre of the sphere.

2 Find the surface area of the solid generated by the rotation of the curve $x = \cos^3 t$, $y = \sin^3 t$ about Ox.

Improper integrals

$\int_a^b f(x)\,dx$ has been defined only when $f(x)$ is defined on the finite closed interval $a \le x \le b$. If, however, $f(x)$ is undefined at some point in this interval or if one (or both) a and b are infinite $\int_a^b f(x)\,dx$ is then called an improper integral. The methods of procedure are illustrated by the following examples.

Examples

1 Evaluate

$$\int_1^\infty \frac{dx}{1+x^2}$$

$$= \lim_{m\to\infty} \int_1^m \frac{dx}{1+x^2}$$

$$= \lim_{m\to\infty} \{[\tan^{-1} x]_1^m\}$$

$$= \lim_{m\to\infty} \{[\tan^{-1} m - \tan^{-1} 1]\}$$

$$= \frac{\pi}{2} - \frac{\pi}{4}$$

$$= \frac{\pi}{4}$$

2 Evaluate

$$\int_0^1 \frac{dx}{\sqrt{1-x^2}}$$

Now $\sqrt{1-x^2} = 0$ when $x = \pm 1$. The integrand is discontinuous at the upper limit of integration $x = 1$.

$$\int_0^1 \frac{dx}{\sqrt{1-x^2}} = \lim_{\varepsilon\to 0^+} \int_0^{1-\varepsilon} \frac{dx}{\sqrt{1-x^2}}$$

$$= \lim_{\varepsilon\to 0^+} \{[\sin^{-1} x]_0^{1-\varepsilon}\}$$

$$= \lim_{\varepsilon\to 0^+} \{\sin^{-1}(1-\varepsilon)\}$$

$$= \sin^{-1} 1 \quad = \frac{\pi}{2}$$

3 Evaluate
$$\int_0^9 (x-1)^{-1/3}\,dx$$

The integrand is discontinuous when $x = 1$.

$$\int_0^9 (x-1)^{1/3}\,dx = \lim_{\varepsilon\to 0^+}\int_0^{1-\varepsilon}(x-1)^{-1/3}\,dx + \lim_{\eta\to 0^+}\int_{1-\eta}^{9}(x-1)^{-1/3}\,dx$$

$$= \lim_{\varepsilon\to 0^+}\left\{\left[\frac{3}{2}(x-1)^{2/3}\right]_0^{1-\varepsilon}\right\} + \lim_{\eta\to 0^+}\left\{\left[\frac{3}{2}(x-1)^{2/3}\right]_{1+\eta}^{9}\right\}$$

$$= \frac{3}{2}\lim_{\varepsilon\to 0^+}\{(-\varepsilon)^{2/3} - (1)\} + \frac{3}{2}\lim_{\eta\to 0^+}\{4 - \eta^{2/3}\}$$

$$= \frac{3}{2}(-1) + \frac{3}{2}(4)$$

$$= \frac{9}{2}$$

N.B. If it is found that a limit obtained in evaluating an improper integral does not exist then that improper integral does not exist.

Practice questions 29

Evaluate the following improper integrals.

1 $\displaystyle\int_{-1}^{1} \frac{dx}{x^{1/3}}$ 2 $\displaystyle\int_0^{\infty} e^{-x}\cos x\,dx$

3 $\displaystyle\int_1^2 \frac{x}{\sqrt{x-1}}\,dx$

3 COMPLEX NUMBERS

The algebra of complex numbers

If a and b are real numbers and $\sqrt{-1}$ is represented by i (or j) then $a + ib$ is called a complex number. Complex numbers obey the usual laws of algebra.

Addition $(a + ib) + (x + iy) = (a + x) + i(b + y)$

Subtraction $(a + ib) - (x + iy) = (a - x) + i(b - y)$

Multiplication
$$(a + ib)(x + iy) = ax + i^2by + ibx + iay$$
$$= (ax - by) + i(bx + ay)$$

Division $\frac{a+ib}{x+iy} = \frac{a+ib}{x+iy} \times \frac{x-iy}{x-iy}$

$$= \frac{(ax+by)+i(bx-ay)}{x^2+y^2}$$

$$= \frac{ax+by}{x^2+y^2} + i\frac{bx-ay}{x^2+y^2}$$

$$i^2 = j^2 = -1$$

Given the complex number $z = a + ib$, a is called the real part of z and b, NOT ib, is called the imaginary part of z. We write

$$\operatorname{Re}(z) = a \qquad \operatorname{Im}(z) = b$$

z is purely real if $b = 0$ and z is purely imaginary if $a = 0$.
The complex conjugate of $z = a + ib$ is the complex number $a - ib$, i.e. the number with the same real part of z and with the sign of the imaginary part of z changed.
If the given complex number is denoted by z the conjugate complex number is denoted by z^* or by $\bar{z}$.

$$z \,.\, z^* = (a+ib)(a-ib)$$
$$= a^2 + b^2\text{, a purely real number.}$$

Example
If $z = 3 - 4i$ and $w = -2 + 5i$, express as complex numbers in the form $a + ib$.

(i) $z + 2w$ (ii) $iz - w$ (iii) zw

(iv) $\frac{iw}{z}$ (v) $\frac{z^*}{(3+i)}w$

(i)
$$z + 2w$$
$$= 3 - 4i + 2(-2 + 5i)$$
$$= 3 - 4i - 4 - 10i$$
$$= \mathbf{-1 - 14i}$$

(ii)
$$iz - w$$
$$= i(3 - 4i) - (-2 + 5i)$$
$$= 3i + 4 + 4 - 5i$$
$$= \mathbf{8 - 2i}$$

(iii)
$$zw = (3 - 4i)(-2 + 5i)$$
$$= -6 + 8i + 15i - 20i^2$$

$$= -6 + 20 + 23i$$
$$= 14 + 23i$$

(iv)
$$\frac{iw}{z} = \frac{i(-2+5i)}{3-4i}$$
$$= \frac{-5-2i}{3-4i} \times \frac{3+4i}{3+4i}$$
$$= \frac{-15-6i-20i-8i^2}{3^2+4^2}$$
$$= \frac{-15+8-26i}{25}$$
$$= -\frac{7}{25} - \frac{26}{25}i$$

(v)
$$\frac{z^*}{(3+i)w}$$
$$= \frac{3+4i}{(3+i)(-2+5i)}$$
$$= \frac{3+4i}{-6-2i+15i+5i^2}$$
$$= \frac{3+4i}{-11+13i}$$
$$= \frac{3+4i}{-11+13i} \times \frac{-11-13i}{-11-13i}$$
$$= \frac{-33-44i-39i-42i^2}{121+169}$$
$$= \frac{9-83i}{290}$$
$$= \frac{9}{290} - \frac{83}{290}i$$

Practice questions 1

1 If $z = 2 + 7i$ and $w = -3 - 4i$, express as complex numbers in the form $a + ib$

(i) $32 + w$ (ii) $2w - iz$ (iii) zw^*

(iv) $\dfrac{z}{iw}$ (v) $\dfrac{(2-i)z^*}{w-z}$

2 Find the real numbers a and b such that

$$\frac{3-i}{-3+4i} - \frac{1-i}{5} = a + ib$$

3 Find real numbers x and y such that

$$\frac{x}{2-3i} - \frac{y}{5+i} = \frac{14+13i}{13(1-i)}$$

Modules and argument or amplitude
Given the complex number $z = x + iy$ we can change its form as follows.

$$\begin{aligned} z &= x + iy \\ &= +\sqrt{x^2+y^2}\left(+\frac{x}{\sqrt{x^2+y^2}} + \frac{iy}{+\sqrt{x^2+y^2}}\right) \end{aligned}$$

Let $r = +\sqrt{x^2+y^2}$, then

$$z = r\left(\frac{x}{r} + i\frac{y}{r}\right)$$

Let θ be such that $\dfrac{x}{r} = \cos\theta$ and $\dfrac{y}{r} = \sin\theta$

then $z = r(\cos\theta + i\sin\theta)$

This is called the **polar form** of the complex number z.

$r = +\sqrt{x^2+y^2}$ is called the **Modulus** of z, written $|z|$.

θ defined by $\cos\theta = x/r$, $\sin\theta = y/r$ (1) is called the **Argument or Amplitude** of z.
Any value of θ satisfying equations (i) is an argument of z, written Arg z. The one value of θ which lies in the interval $-\pi < \theta \leq \pi$ is called the principal value of the Argument, written arg z.

Arg z = arg $z + 2\pi n$ when n is a positive or negative integer. The angle is normally given in radians.

Example
Find the modulus and the argument of the complex number $z = \sqrt{3} - i$.

$$\begin{aligned} |z| &= \sqrt{3+1} \\ &= 2 \end{aligned}$$

$$z = \sqrt{3} - i$$

$$= 2\left(\frac{\sqrt{3}}{2} - i\frac{1}{2}\right)$$

$$\cos\theta = \frac{\sqrt{3}}{2} \quad \sin\theta = -\frac{1}{2} \quad \text{(i)}$$

One value of θ which satisfies equation (i) is $11\pi/6$, (330°)
The principal value of the argument is therefore

$-\frac{\pi}{6}$, (−30°)

[This could have been obtained directly from equation (i)]

$$\arg z = -\frac{\pi}{6}$$

$$\text{Arg}\, z = -\frac{\pi}{6} + 2\pi n, \; n \text{ an integer.}$$

Practice questions 2

1 Find the modulus and principal value of the argument of each of the following complex numbers.

(a) $1 - i$ (b) $+4\sqrt{3} + 4i$ (c) $-\sqrt{3} + i$

2 Express in modulus-argument from the complex numbers

(a) $2 + i$ (b) $\dfrac{1 - i}{1 + i}$

Multiplication and division of complex numbers in polar forms
If $z_1 = r_1(\cos\theta_1 + i\sin\theta_1)$ and $z_2 = r_2(\cos\theta_2 + i\sin\theta_2)$

$$\begin{aligned}
z_1 z_2 &= r_1 r_2(\cos\theta_1 + i\sin\theta_1)(\cos\theta_2 + i\sin\theta_2)\\
&= r_1 r_2(\cos\theta_1\cos\theta_2 - \sin\theta_1\sin\theta_2 + i\sin\theta_1\cos\theta_2 + i\cos\theta_1\sin\theta_2)\\
&= r_1 r_2[\cos(\theta_1 + \theta_2) + i\sin(\theta_1 + \theta_2)]
\end{aligned}$$

$$\therefore \; |z_1 z_2| = |z_1||z_2|$$

and $\text{Arg}\,(z_1 z_2) = \text{Arg}\, z_1 + \text{Arg}\, z_2$

$$\frac{z_1}{z_2} = \frac{r_1(\cos\theta_1 + i\sin\theta_1)}{r_2(\cos\theta_2 + i\sin\theta_2)}$$

$$= \frac{r_1}{r_2}\,\frac{(\cos\theta_1 + i\sin\theta_1)(\cos\theta_2 - i\sin\theta_2)}{(\cos\theta_2 + i\sin\theta_2)(\cos\theta_2 - i\sin\theta_2)}$$

$$= \frac{r_1}{r_2}[\cos(\theta_1 - \theta_2) + i\sin(\theta_1 - \theta_2)]$$

$$\therefore \left|\frac{z_1}{z_2}\right| = \left|\frac{z_1}{z_2}\right|$$

and $\operatorname{Arg}\left(\frac{z_1}{z_2}\right) = \operatorname{Arg} z_1 - \operatorname{Arg} z_2$

Examples

1 Express as a single complex number in polar form

$3(\cos 4\theta - i\sin 4\theta) \times 5(\cos 2\theta + i\sin 2\theta)$
$z = 3(\cos 4\theta - i\sin 4\theta) \times 5(\cos 2\theta + i\sin 2\theta)$

Now $\cos 4\theta = \cos(-4\theta)$ and $-\sin 4\theta = \sin(-4\theta)$

$$\therefore\ z = 3[\cos(-4\theta) + i\sin(-4\theta)] \times 5(\cos 2\theta + i\sin 2\theta)$$
$$= 15[\cos(-4\theta + 2\theta) + i\sin(-4\theta + 2\theta)]$$
$$= \mathbf{15[\cos(-2\theta) + i\sin(-2\theta)]}$$

2 Express as single complex numbers in polar form

$$z = \frac{3(\cos 8\alpha + i\sin 8\alpha)}{4(\cos 2\alpha - i\sin 2\alpha)} \quad \text{and } w = \frac{\left(\cos\frac{3\pi}{2} + i\sin\frac{3\pi^2}{2}\right)^2}{-8i}$$

$$z = \frac{3(\cos 8\alpha + i\sin 8\alpha)}{4(\cos 2\alpha - i\sin 2\alpha)}$$
$$= \frac{3}{4}\,\frac{(\cos 8\alpha + i\sin 8\alpha)}{[\cos(-2\alpha) + i\sin(-2\alpha)]}$$
$$= \frac{3}{4}[\cos(8\alpha + 2\alpha) + i\sin(8\alpha + 2\alpha)]$$
$$= \mathbf{\frac{3}{4}[\cos 10\alpha + i\sin 10\alpha]}$$

$$w = \frac{\left(\cos\frac{3\pi}{2} + i\sin\frac{3\pi}{2}\right)^2}{-8i}$$
$$= \frac{\cos 3\pi + i\sin 3\pi}{8\left[\cos\left(-\frac{\pi}{2}\right) + i\sin\left(-\frac{\pi}{2}\right)\right]}$$
$$= \frac{1}{8}\left[\cos\frac{7\pi}{2} + i\sin\frac{7\pi}{2}\right]$$

$$= \frac{1}{8}\left[\cos\left(-\frac{\pi}{2}\right) + i\sin\left(-\frac{\pi}{2}\right)\right],$$

giving the principal value of the argument.

Practice questions 3

Express the following as single complex numbers in polar form.

1 $2(\cos\alpha + i\sin\alpha) \times 3(\cos 3\alpha + i\sin 3\alpha)$

2 $5(\cos 4\theta - i\sin 4\theta)$

3 $\dfrac{1}{\cos\alpha - i\sin\alpha}$

4 $\dfrac{2(\cos 6\alpha + i\sin 6\alpha)}{3(\cos 2\alpha + i\sin 2\alpha)}$

5 $i\left(\cos\dfrac{\pi}{6} - i\sin\dfrac{\pi}{6}\right)^2$

6 $\dfrac{\cos\dfrac{3\pi}{4} - i\sin\dfrac{3\pi}{4}}{-2i}$

7 $1 + i\tan\theta$

Geometrical representation of complex numbers

The Argand diagram

The complex number $z = x + iy$ can be represented uniquely by the point P with co-ordinates (x, y) in a Cartesian plane. If the origin O is taken as pole and Ox as the initial line then the polar coordinates of $P(r, \theta)$ are such that

$$r = \sqrt{x^2 + y^2} \quad \text{and} \quad \cos\theta = \frac{x}{r}, \quad \sin\theta = \frac{y}{r}$$

i.e. the polar coordinates of P defined as above give the modulus and argument of the complex number $z = r(\cos\theta + i\sin\theta)$. This is sometimes written as $r \operatorname{cis} \theta$. Again the vector $\overline{OP}$ may be regarded as a representation of the complex number z since the length of $\overline{OP}$ gives the modulus of z, and the angle which $\overline{OP}$ makes with Ox gives the argument of z.

P is called the affix of z.

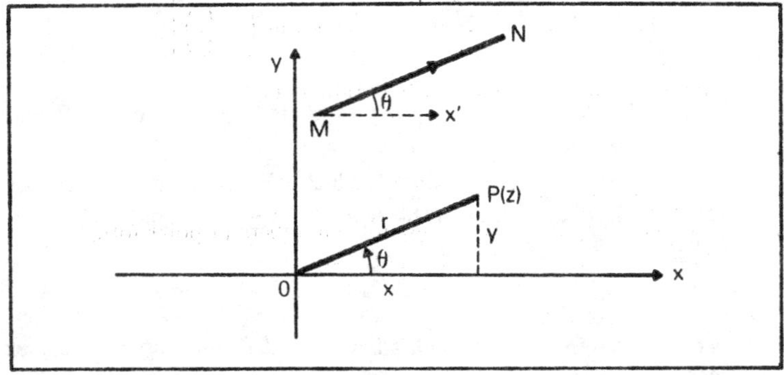

If $\overline{MN} = \overline{OP}$ then $\overline{MN}$ also represents the complex number z, since length of $\overline{MN}$ = length of $\overline{OP} = |z|$ and $\angle POx = \angle NMx' = \theta = \text{Arg}\, z$. What MN does not give is the affix of z.

$\overline{MN}$ gives the polar form of z directly. It does not give the $x + iy$ form directly.

Geometrical representation of addition and subtraction of two complex numbers

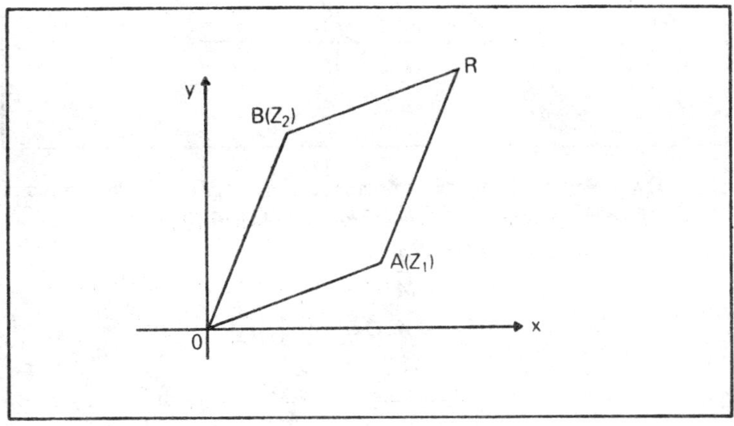

A is the affix of $z_1 = x_1 + iy_1$ and B is the affix of $z_2 = x_2 + iy_2$. i.e. the coordinates of A and B are (x_1, y_1) and (x_2, y_2) respectively.

$$z_1 + z_2 = (x_1 + x_2) + i(y + y_2)$$

Completing the parallelogram $O\,A\,R\,B$ it is seen that the coordinates of R are $[(x_1 + x_2), (y_1 + y_2)]$ therefore R is the affix of $(z_1 + z_2)$ and $\overline{OR}$ represents $z_1 + z_2$.

Again $\overline{OB} = \overline{OA} + \overline{AB}$

$\therefore\ \overline{OB} - \overline{OA} = \overline{AB}$

i.e. the complex number $z_2 - z_1$ is represented by the vector $\overline{AB}$ and $|z_2 - z_1|$ is represented by the length of $\overline{AB}$.
Arg $(z_2 - z_1)$ is represented by the angle which $\overline{AB}$ makes with Ox. In order to obtain the affix of $z_2 - z_1$ draw $\overline{OM}$ from the origin such that $\overline{OM} = \overline{AB}$. Then M is the affix of $z_2 - z_1$.

Geometrical representation of multiplication and division of two complex numbers

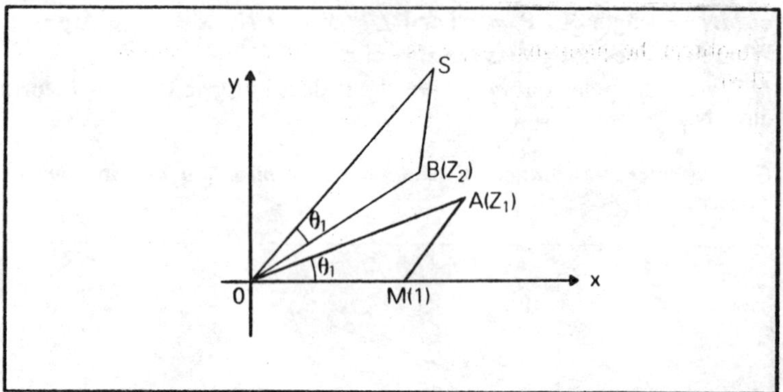

A and B represent the complex numbers z_1 and z_2 respectively. M represents the complex number 1. $\triangle\, O\, B\, S$ is drawn directly similar to $\triangle\, O\, M\, A$, then

$$\frac{OS}{OB} = \frac{OA}{OM}$$

$$OS = O\,A\,.\,O\,B = |z_1|\,.\,|z_2|$$

Also

$$\angle SOx = \angle SOB + \angle BOx$$
$$= \angle AOM + \angle BOx$$
$$= \text{Arg}\, z_1 + \text{Arg}\, z_2$$

$\therefore\ \overline{OS}$ represents $z_1 z_2$

S is the affix of the product z_1z_2

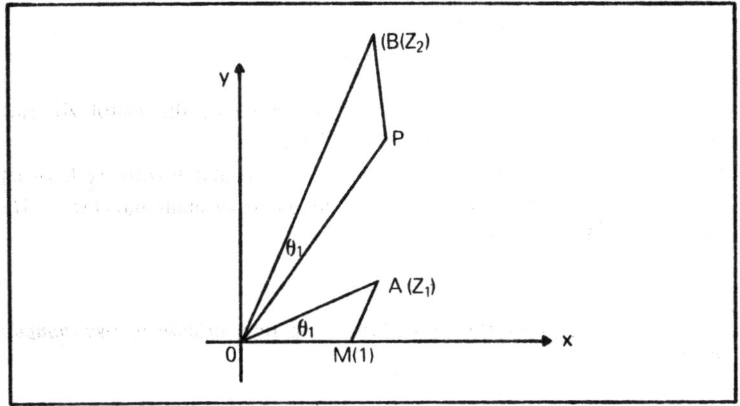

To obtain the quotients z_2/z_1 construct $\triangle OBP$ directly similar to $\triangle OAM$, then

$$\frac{OP}{OB} = \frac{1}{OA}$$

$$OP = \frac{OB}{OA}$$

$$= \frac{|z_2|}{|z_1|}$$

$$\angle POx = \angle BOx - \angle BOP$$

$$= \angle BOx - \angle AOM$$

$$\angle POx = \angle BO_x - \angle BOP$$

$$= \angle BO_x - \angle AOM$$

$$= = \text{Arg}\, z_2 - \text{Arg}\, z_1$$

$$\therefore\ \overline{OP} \text{ represents } \frac{z_2}{z_1}$$

P is the affix of the quotient $\frac{z_2}{z_1}$

Example
The complex numbers $z_1 = 4 + i$ and $z_2 = 2 + 3i$ are represented in an Argand diagram by the points A and B respectively. ABC and ABD are equilateral triangles constructed on AB as base. Find the complex numbers represented by the points C and D.

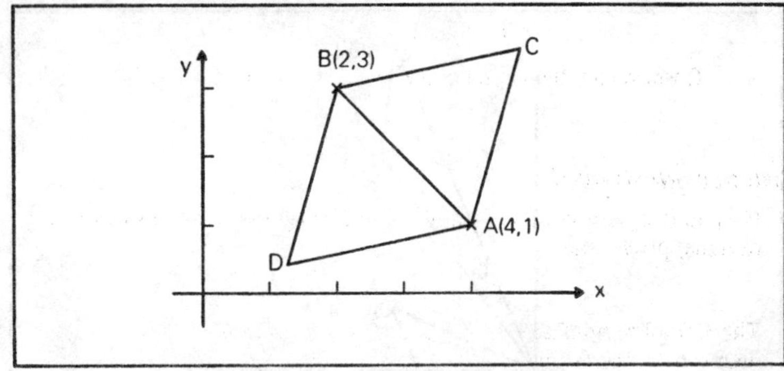

$\overline{AB} = \overline{OB} - \overline{OA}$

$\overline{AB}$ represents the complex number $z_2 - z_1 = -2 + 2i$

To rotate $\overline{AB}$ through an angle of $\pi/3$ (to coincide with $\overline{AD}$) we require a complex number z such that $|z| = 1$ and $\arg z = \pi/3$.

$$\therefore\ z = 1\left(\cos\frac{\pi}{3} + i\sin\frac{\pi}{3}\right)$$

$$= \tfrac{1}{2}(1 + i\sqrt{3})$$

Then $\overline{AD}$ represents $\frac{1}{2}(1 + i\sqrt{3})(-2 + 2i)$

$$= -1 - \sqrt{3} + i(1 - \sqrt{3})$$

$\overline{OD} = \overline{OA} + \overline{AD}$

$\overline{OD}$ represents $4 + i + [-1 - \sqrt{3} + i(1 - \sqrt{3})]$

$$= 3 - \sqrt{3} + i(2 - \sqrt{3})$$

D represents the complex number $3 - \sqrt{3} + i(2 - \sqrt{3})$

To rotate $\overline{AB}$ through an angle $-\dfrac{\pi}{3}$ (to coincide with $\overline{AC}$) we require a complex number w such that $|w| = 1$ and $\arg w = -\pi/3$.

$$\therefore\ w = 1\left[\cos\left(-\frac{\pi}{3}\right) + i\sin\left(-\frac{\pi}{3}\right)\right] = \tfrac{1}{2}(1 - i\sqrt{3})$$

Then $\overline{AC}$ represents $\frac{1}{2}(1 - i\sqrt{3})(-2 + 2i) = -1 + \sqrt{3} + i(1 + \sqrt{3})$

$\overline{OC} = \overline{OA} + \overline{AC}$

$\overline{OC}$ represents $4 + i + [-1 + \sqrt{3} + i(1 + \sqrt{3})]$

$$= 3 + \sqrt{3} + i(2 + \sqrt{3})$$

C represents the complex number $3 + \sqrt{3} + i(2 + \sqrt{3})$

Practice questions 4

1 If z_1 and z_2 are complex numbers with arguments differing by $\pi/2$ radians, prove that

$$|z_1 - z_2| = |z_1 + z_2|$$

2 The complex numbers $4 - i$ and 2 are represented in an Argand diagram by the points A and B respectively.
$ABCD$ and ABC_1D_1 are squares on AB as base. Find the complex numbers represented by the points C, D, C_1 and D_1.

3 In an Argand diagram ABC is an equilateral triangle whose circumcentre is the origin. If A represents the complex number $3 + 4i$, find the complex numbers represented by B and C.

4 If z_1 and z_2 are complex numbers such that

$$\text{Arg}\left(\frac{z_1 + z_2}{z_1 - z_2}\right) = \frac{\pi}{2} \quad \text{prove that}$$

$$|z_1| = |z_2|$$

Complex numbers and loci

The following examples illustrate the various techniques which may be used.

Examples

1 If z is a complex number and $|z - 1 + 2i| = 3$, find the locus of the point which represents z in the Argand diagram.

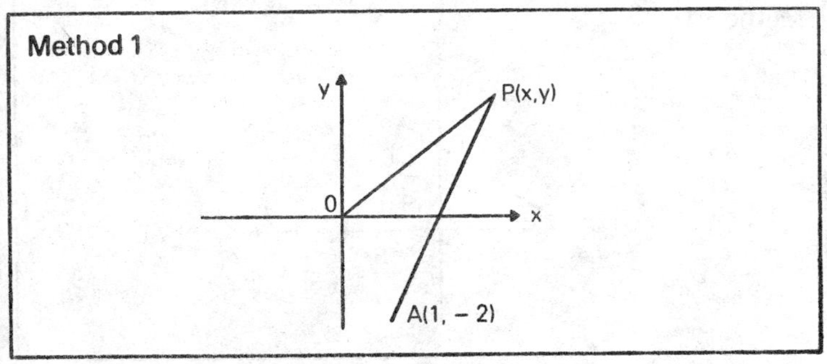

$$z - 1 + 2i = z - (1 - 2i)$$

Let $P(x, y)$ represent the complex no. z and let $A(1, -2)$ represent $1 - 2i$

$$\overline{AP} = \overline{OP} - \overline{OA}$$

$\therefore$ $\overline{AP}$ represents the complex number $z - (1 - 2i)$

But $|z - 1 + 2i| = 3$ and A is fixed

$\therefore$ **the locus of P is a circle centre A (1, −2) with radius 3**

N.B. It is easier to represent the difference of two complex numbers in the Argand diagram than to represent their sum so $z - 1 + 2i$ was written in the form $z - (1 - 2i)$.

Method 2

Let $P(x, y)$ represent $z = x + iy$

$$\text{Then } z - 1 + 2i = x + iy - 1 + 2i$$
$$= (x - 1) + i(y + 2)$$
$$|z - 1 + 2i|^2 = (x - 1)^2 + (y + 2)^2$$

But $|z - 1 + 2i| = 3$, given

$$\therefore (x - 1)^2 + (y + 2)^2 = 3$$

which is the equation of a circle centre (1, −2) radius 3

2 Find the locus of the point P which represents the complex number z in an Argand diagram if

$$\arg\frac{z - 2}{z + 2} = \frac{\pi}{4}$$

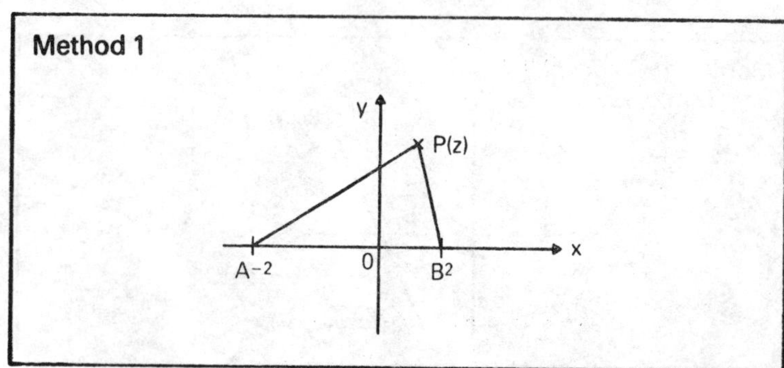

If A is the point $(-2, 0)$ and B the point $(2, 0)$

$$z + 2 \text{ is represented by } \overline{AP}$$

$$z - 2 \text{ is represented by } \overline{BP}$$

$$\arg \frac{z-2}{z+2} \text{ is represented by } \angle BPA$$

$$\therefore \angle BPA = \frac{\pi}{4}$$

i.e. P moves on the arc of a circle with AB as chord.
The centre of this circle must be on Oy at C say. Then $\angle BCA = \pi/2$ ($\angle$ at centre and circumference).
$\triangle ACB$ is therefore a right-angled isosceles triangle.
$\therefore$ **C is the point (0, 2) and the radius of the circle is $2\sqrt{2}$.**

Method 2
Let $P(x, y)$ represent $z = x + iy$

$$\frac{z-2}{z+2} = \frac{x-2+iy}{x+2+iy}$$

$$= \frac{[(x-2)+iy][(x+2)-iy]}{(x+2)^2+y^2}$$

$$= \frac{x^2-4+y^2+4iy}{(x+2)^2+y^2}$$

$$= u + iv \text{ say}$$

$$\arg \frac{z-2}{z+2} = \tan^{-1}\frac{v}{u}$$

$$\text{But } \arg \frac{z-2}{z+2} = \frac{\pi}{4}$$

$$\therefore \frac{v}{u} = \tan\frac{\pi}{4}$$

$$\frac{4y}{x^2-4+y^2} = 1$$

$$x^2 - 4 + y^2 = 4y$$

$$x^2 + y^2 - 4y + 2^2 - 4 - 4 = 0$$

$$x^2 + (y-2)^2 - 8 = 0$$

which is the equation of a circle centre (2, 0) and radius $2\sqrt{2}$.

3 If $z = x + iy$ and the ratio $(z + i/z - i)$ is purely imaginary, find the locus of the affix of z.

$$\text{Let } W = \frac{z - 2 + i}{z - i}$$

$$= \frac{x - 2 + i(y + 1)}{x + i(y - 1)}$$

$$= \frac{[(x - 2) + i(y + 1)][x - i(y - 1)]}{x^2 + (y - 1)^2}$$

Since W is purely imaginary real $W = 0$

$$\text{i.e. } \frac{x(x - 2) + (y + 1)(y - 1)}{x^2 + (y - 1)^2} = 0$$

$$\therefore\ x^2 - 2x + y^2 - 1 = 0$$

$$(x - 1)^2 + y^2 = 2$$

the locus of the affix of z is a circle centre (1, 0) radius 2.

4 Find the locus of $P(x, y)$ if $z = x + iy$ is such that $\arg(z - 1 - i) = \pi/3$

Method 1

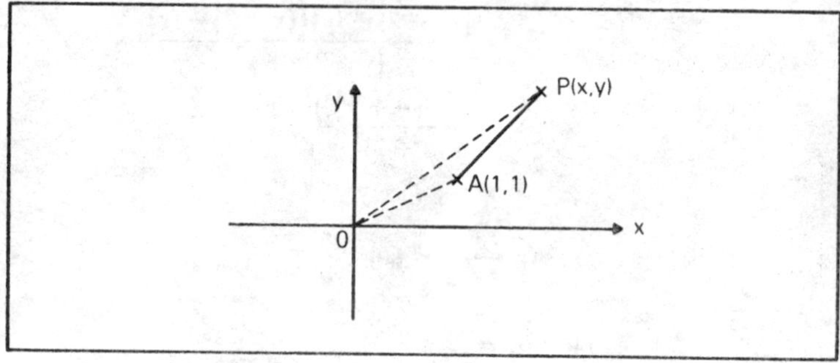

Let A (1, 1) represent $1 + i$
Then $\overline{AP}$ represents $z - 1 - i$

$$\arg(z - 1 - i) = \angle PAx = \frac{\pi}{3}$$

i.e. the angle which $\overline{AP}$ makes with Ox is $\frac{\pi}{3}$

$\therefore$ locus of P is that part of the straight line passing through A (1, 1) making an angle π with Ox. (i.e. the part of the line above A)

Method 2

$$z = x + iy$$

$$\therefore\ z - 1 - i = x - 1 + i(y - 1)$$

$$\arg(z - 1 - i) = \tan^{-1}\frac{y-1}{x-1}$$

$$= \frac{\pi}{3}$$

$$\therefore\ \frac{y-1}{x-1} = \tan\frac{\pi}{3}$$

$$\frac{y-1}{x-1} = \sqrt{3}$$

$$y - 1 = \sqrt{3}x - \sqrt{3}$$

$$y = \sqrt{3}x + 1 - \sqrt{3}$$

the locus of P is the straight line with equation y = $\sqrt{3x} + 1 - \sqrt{3}$

N.B. This method does not distinguish between the two parts of this line and so does not give the complete answer.

5 If $P(x_1y)$ represents the complex number z, describe the region of the Argand diagram represented by

$$|z - 2 - 3i| \leq 3$$

Hence prove that

$$3\sqrt{2} - 3 \leq |z - 5| \leq 3\sqrt{2} + 3$$

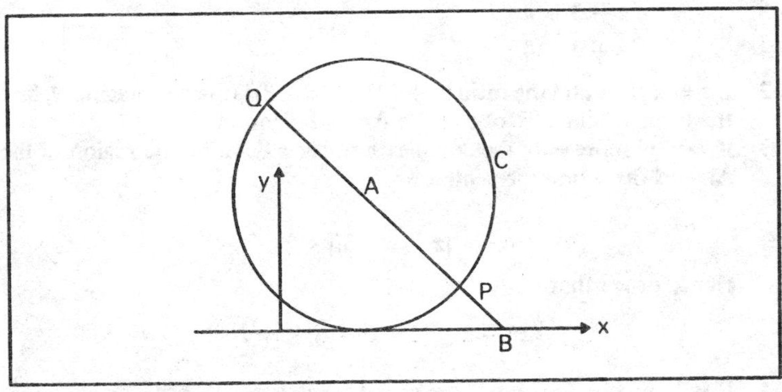

A is the point (2, 3)
Then $z - 2 - 3i$ is represented by $\overline{AP}$
and since $|z - 2 - 3i| \leq 3$, $AP \leq 3$

the region represented is that on and inside the circle centre $A(2, 3)$ and radius 3.

Consider $|z - 5| = k$ a real constant.
This represents a circle centre $B(5, 0)$ radius k.
To satisfy the given condition, this circle must intersect the circle centre $A(2, 3)$ radius 3, circle C, say. If the line BA joining the centres meets the circle C in P and Q then k, the radius of the second circle, must lie between BP and BQ.

$$AB^2 = (2 - 5)^2 + (3 - 0)^2$$
$$= 18$$
$$AB = 3\sqrt{2}$$
$$BP = 3\sqrt{2} - 3$$
$$BQ = 3\sqrt{2} + 3$$
$$\therefore\ 3\sqrt{2} - 3 \leq |z - 5| \leq 3\sqrt{2} + 3$$

Practice questions 5

1 If z is a complex number find the locus of the point which represents z in the Argand diagram, when

(i) $|z - 1 + 5i| = 6$

(ii) $\text{Arg}(z - 2) = \dfrac{\pi}{4}$

(iii) $\text{Arg}\dfrac{z - 3}{z + 3} = \dfrac{\pi}{4}$

2 If $z = x + iy$ and the ratio $(z + 2 + i : (z - 3)$ is purely imaginary, find the locus of the affix of z in the Argand diagram.

3 If $P(x, y)$ represents the complex number z describe the region of the Argand diagram represented by

$$|z - 1 - 3i| \leq 2$$

Hence prove that

$$\sqrt{26} - 2 \leq |z - 6 - 2i| \leq \sqrt{26} + 2$$

4 If $z = x + iy$ find the locus of the point $P(x, y)$ when

$$\left|\frac{z - 2}{z + 2}\right| = 3$$

Transformations and complex numbers

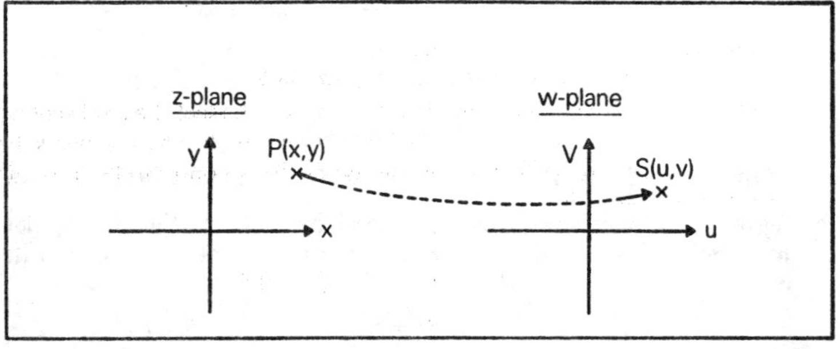

Let $w = f(z)$ be a single valued function of $z = x + iy$ in a given region R of the z-plane. Thus every point $P(x, y)$ of R is mapped onto a unique point $S(u, v)$ in the w-plane where $w = u + iv$. If, for example, $w = 2z + 3i$, the point P representing $z = -1 + 3i$ is mapped onto the point $S(u, v)$ given by

$$u + iv = 2(-1 + 3i) + 3i$$

$$u + iv = -2 + 9i$$

$$\therefore \; S \text{ is the point } (-2, 9)$$

N.B. Do not try to superimpose the image points S on the z-plane. Always consider two separate planes as shown above.

Examples

1 Under the mapping $w = \dfrac{1}{z}$ find the locus of the point w when the locus of z is the circle whose equation is $x^2 + y^2 + 2x + 4y = 0$.
We have to convert from an equation in x and y to an equation in u and v i.e. from z to w.

$$w = \frac{1}{z}$$

$$\therefore \; z = \frac{1}{w}$$

$$x + iy = \frac{1}{u + iv}$$

$$= \frac{u - iv}{u^2 + v^2}$$

equating real and imaginary parts

$$x = \frac{u}{u^2 + v^2} \quad \text{and} \quad y = -\frac{v}{u^2 + v^2}$$

$x^2 + y^2 + 2x + 4y = 0$ becomes

$$\left(\frac{u}{(u^2+v^2)}\right)^2 + \left(-\frac{v}{u^2+v^2}\right)^2 + 2\left(\frac{u}{u^2+v^2}\right) + 4\left(-\frac{v}{u^2+v^2}\right) = 0$$

$$\frac{1}{u^2+v^2} + \frac{2u}{u^2+v^2} - \frac{4v}{u^2+v^2} = 0$$

$$1 + 2u - 4v = 0$$

or **2u − 4v + 1 = 0 which is the equation of a straight line in the w-plane.**

2 Show that under the mapping defined by $w = z^2$ the rectangular hyperbola $x^2 - y^2 = 4$ transforms into a straight line and find its equation.

$$w = z^2$$

In this case $z = w^{1/2}$ and $x + iy = (u + iv)^{1/2}$ which does not lead us directly to obtaining x and y as functions of u and v. We proceed as follows

$$w = z^2$$

$$\therefore\; u + iv = (x + iy)^2$$

$$u + iv = x^2 - y^2 + 2ixy$$

Equating real and imaginary parts we obtain

$$u = x^2 - y^2 \quad \text{and} \quad v = 2xy$$

Now the given locus is $x^2 - y^2 = 4$
Hence the image of this locus is

u = 4, a straight line in the w-plane

Practice questions 6

1 Under the mapping defined by $w = \dfrac{1}{z}$ find the image locus of the circle $x^2 + y^2 + 3x = 0$.

2 Show that under the mapping $w = i/(z + 1)$ the circle $|z + 1| = 1$ is mapped on to the circle $|w| = 1$. Find the image region of the region $|z + 1| < 1$.

3 Find the image of the circle $|z| = 2$ under the mapping $w = (z + i)/(z - 1)$.

Applications of De Moivre's theorem

De Moivre's theorem states

(i) for any positive or negative integer n.
$(\cos\theta + i\sin\theta)^n = \cos n\theta + i\sin n\theta$.

(ii) for any rational number p/q where p, q are integers and $q > 0$,

$$\cos\frac{p}{q}\theta + i\sin\frac{p}{q}\theta \text{ is a value of } (\cos\theta + i\sin\theta)^{p/q}.$$

Examples

1 Express as a single complex number in $a + ib$ form

(a) $(1 + i)^7$ (b) $(-1 + i\sqrt{3})^6$

(a) We first express $1 + i$ in polar form. A diagram often helps to clarify the value of the modulus and argument.

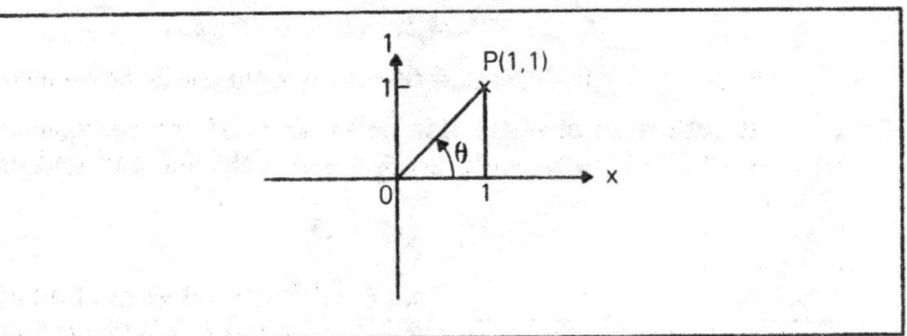

$$|1 + i| = OP = \sqrt{2}$$

$$\arg(1 + i) = \angle POx = \frac{\pi}{4}$$

$$1 + i = \sqrt{2}\left(\cos\frac{\pi}{4} + i\sin\frac{\pi}{4}\right)$$

$$(1 + i)^7 = (\sqrt{2})^7\left(\cos\frac{\pi}{4} + i\sin\frac{\pi}{4}\right)^7$$

$$= (\sqrt{2})^7\left(\cos\frac{7\pi}{4} + i\sin\frac{7\pi}{4}\right)$$

$$= (\sqrt{2})^7\left(\frac{1}{\sqrt{2}} - i\frac{1}{\sqrt{2}}\right)$$

$$= 8(1 - i)$$

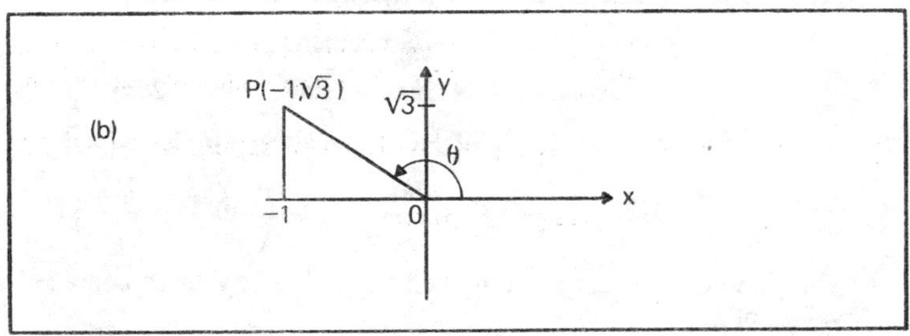

$$|-1 + i\sqrt{3}| = OP = 2$$

$$\arg(-1 + i\sqrt{3}) = \angle\theta = \frac{2\pi}{3}$$

$$-1+\sqrt{3}=2\left(\cos\frac{2\pi}{3}+i\sin\frac{2\pi}{3}\right)$$

$$(-1+i\sqrt{3})^6=2^6\left(\cos\frac{2\pi}{3}+i\sin\frac{2\pi}{3}\right)^6$$

$$=2^6(\cos 4\pi+i\sin 4\pi)$$

$$=64$$

2 Find the cube roots of $-2-3i$ in polar form.

$$z=-2-3i$$

$$|z|=\sqrt{2^2+3^2}$$

$$=\sqrt{13}$$

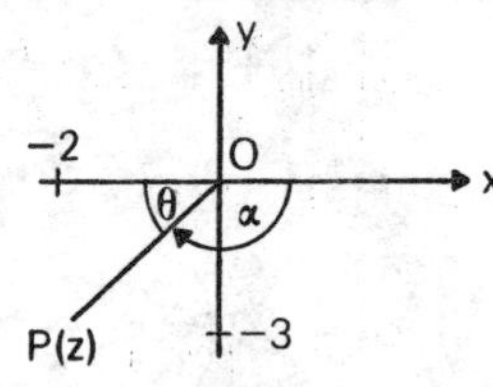

principal value of $\arg z=-\alpha$ where α is obtuse

and $\sin\alpha=\dfrac{3}{\sqrt{13}}$, $\cos\alpha=\dfrac{-2}{\sqrt{13}}$

i.e. α is obtuse and $\tan\alpha=-\dfrac{3}{2}$

$$\alpha=56{\cdot}3°=0{\cdot}99 \text{ radians}$$

the general value of $\arg z=-\alpha+2k\pi$ where k is an integer

$$=0{\cdot}99+2k\pi \text{ radians}$$

$$z=13^{1/2}[\cos(-0{\cdot}99+2k\pi)+i\sin(-0{\cdot}99+2k\pi)]$$

$$z^{1/3}=13^{1/6}[\cos\tfrac{1}{3}(-0{\cdot}99+2k\pi)+i\sin\tfrac{1}{3}(-0{\cdot}99+2k\pi)]$$

$$=13^{1/6}\left[\cos\left(-0{\cdot}33+\frac{2k\pi}{3}\right)+i\sin\left(-0{\cdot}33+\frac{2k\pi}{3}\right)\right]$$

The three cube roots are obtained by taking any three consecutive values of k.

when $k=0$, $z_1^{1/3}=13^{1/6}[\text{cis}(-0{\cdot}33)]$

when $k=1$, $z_2^{1/3}=13^{1/6}\left[\text{cis}\left(-0.33+\dfrac{2\pi}{3}\right)\right]$

when $k = 2$, $z_3^{1/3} = 13^{1/6}\left[\text{cis}\left(-0.33 + \frac{4\pi}{3}\right)\right]$.

3 Solve the equation $(z - 2i)^6 = 64$

$$(z - 2i)^6 = 64$$
$$= 2^6$$

expressing 2^6 as a complex number in polar form giving the general expression for its argument

$$(z - 2i)^6 = 2^6[\cos(0 + 2k\pi) + i\sin(0 + 2k\pi)]$$

where k is an integer

$$z - 2i = 2\left[\cos\frac{2k\pi}{6} + i\sin\frac{2k\pi}{6}\right]$$
$$= 2\left[\cos\frac{k\pi}{3} + i\sin\frac{k\pi}{3}\right]$$

when $k = 0$, $z - 2i = 2(\cos 0 + i\sin 0)$

$$= 2$$
$$\mathbf{z = 2 + 2i}$$

when $k = 1$, $z - 2i = 2\left(\cos\frac{\pi}{3} + i\sin\frac{\pi}{3}\right)$

$$= 2\left(\frac{1}{2} + i\frac{\sqrt{3}}{2}\right)$$
$$= 1 + i\sqrt{3}$$
$$\mathbf{z = 1 + i(\sqrt{3} + 2)}$$

$k = 2$, $z - 2i = 2\left(\cos\frac{2\pi}{3} + i\sin\frac{2\pi}{3}\right)$

$$= 2\left(-\frac{1}{2} + i\frac{\sqrt{3}}{2}\right)$$
$$= -1 + i\sqrt{3}$$
$$\mathbf{z = -1 + i(\sqrt{3} + 2)}$$

$k = 3$, $z - 2i = 2(\cos\pi + i\sin\pi)$

$$= -2$$
$$\mathbf{z = -2 + 2i}$$

$k = 4,$
$$z - 2i = 2\left(\cos\frac{4\pi}{3} + i\sin\frac{4\pi}{3}\right)$$
$$= 2\left(-\frac{1}{2} - i\frac{\sqrt{3}}{2}\right)$$
$$= -1 - i\sqrt{3}$$
$$\mathbf{z = -1 + i(2 - \sqrt{3})}$$

$k = 5,$
$$z - 2i = 2\left(\cos\frac{5\pi}{3} + i\sin\frac{5\pi}{3}\right)$$
$$= 2\left(\frac{1}{2} - i\frac{\sqrt{3}}{2}\right)$$
$$= 1 - i\sqrt{3}$$
$$\mathbf{z = 1 + i(2 - \sqrt{3})}$$

N.B. when $k = 6$,
$$z - zi = 2(\cos 2\pi + i\sin 2\pi)$$
$$= 2(\cos 0 + i\sin 0)$$
which gives the same value of z as that obtained by taking $k = 0$. There are only 6 distinct roots.
The roots of the given equation are thus
$$\mathbf{\pm 1 + i(2 + \sqrt{3}),\quad \pm 1 + i(2 - \sqrt{3}),\quad \pm 2 + 2i}$$

4 Show that the nth roots of 1 (a) form a geometric progression, (b) the non-real roots occur in conjugate pairs.

Let $z^n = 1$
$$= \cos 2k\pi + i\sin k\pi,\ k \text{ an integer}$$
$$z = \sqrt[n]{1}$$
$$= \cos\frac{2k\pi}{n} + i\sin\frac{-2k\pi}{n},\ k = 0, 1, 2\ldots(n-1)$$

the nth roots of 1 are $1, \operatorname{cis}\frac{2\pi}{n}, \operatorname{cis}\frac{4\pi}{n}, \operatorname{cis}\frac{6\pi}{n} \ldots \operatorname{cis}\frac{2(n-1)\pi}{n}$

i.e. the nth roots of 1 are $1, \operatorname{cis}\frac{2\pi}{n}, \left(\operatorname{cis}\frac{2\pi}{n}\right)^2, \left(\operatorname{cis}\frac{2\pi}{n}\right)^3 \ldots \left(\operatorname{cis}\frac{2\pi}{n}\right)^{n-1}$

which shows that the nth roots of 1 form a G.P. with first term 1 and common ratio cis $2\pi/n$

The last root, $\text{cis}\,\dfrac{2(n-1)\pi}{n}$

$$= \cos\frac{2(n-1)\pi}{n} + i\sin\frac{2(n-1)\pi}{n}$$

$$= \cos\frac{2\pi}{n} - i\sin\frac{2\pi}{n}$$

which is the complex conjugate of the first complex root

$$\cos\frac{2\pi}{n} + i\sin\frac{2\pi}{n}$$

Similarly we can pair off the other complex roots in this way.
When n is odd there will be $(n-1)/2$ pairs of complex roots plus the real root 1.
When n is even there will be $(n-2)/2$ pairs of complex roots plus the real roots 1 and -1 since $-1 = \text{cis}\,(n\pi)/n$.
Hence when n is odd

$$z = 1 \text{ and } z = \text{cis}\left(\pm\frac{2k\pi}{n}\right)$$

$$= \cos\frac{2k\pi}{n} \pm i\sin\frac{2k\pi}{w}, \quad k = 1, 2, 3, \ldots, \frac{n-1}{2}$$

and when n is even

$$z = \pm 1 \text{ and } z = \text{cis}\left(\pm\frac{2k\pi}{n}\right)$$

$$= \cos\frac{2k\pi}{n} \pm i\sin\frac{2k\pi}{n}, \quad k = 1, 2, 3, \ldots, \frac{n-2}{2}$$

Practice questions 7

1 Express as a single complex number in $a + ib$ form

(a) $(-1+i)^5$ (b) $(-1-i\sqrt{3})^8$ (c) $(\sqrt{3}-i)^6$

2 Find the 4th roots of $2 - i$ in $a + ib$ form giving the real and the imaginary part of each root correct to 2 decimal places.

3 Find the roots of the equation

$$(z+1)^4 = 81$$

4 Find the nth roots of -1, when n is even and greater than 2 and hence express $x^n + 1$ as the product of real quadratic factors.

$\cos^n \theta$ and $\sin^n \theta$ in terms of cosine and sine of multiples of θ and vice versa

If $\quad z = \cos\theta + i\sin\theta \quad$ then $\quad z^n = \cos n\theta + i\sin n\theta$,

$z^{-1} = \cos\theta - i\sin\theta \quad$ and $\quad z^{-n} = \cos n\theta - i\sin n\theta$

$$\left.\begin{aligned} \therefore\ z + \frac{1}{z} &= 2\cos\theta \\ \text{and } z - \frac{1}{z} &= 2i\sin\theta \end{aligned}\right\} \text{(A)}$$

$$\left.\begin{aligned} z^n + \frac{1}{z^n}n &= 2\cos n\theta \\ \text{and } z^n - \frac{1}{z^n} &= 2i\sin n\theta \end{aligned}\right\} \text{(B)}$$

Results A and B enable positive integral powers of $\cos\theta$ and $\sin\theta$ to be expressed in terms of cosines and sines of multiples of θ.

Examples

1 Express $\sin^5\theta$ in terms of sines of multiples of θ.
Let $z = \cos\theta + i\sin\theta$

Then $z - \dfrac{1}{z} = 2i\sin\theta$

$$(2i\sin\theta)^5 = \left(z - \frac{1}{z}\right)^5$$

$$= z^5 - 5z^3 + 10z - \frac{10}{z} + \frac{5}{z^3} - \frac{1}{z^5}$$

$$= z^5 - \frac{1}{z^5} - 5\left(z^3 - \frac{1}{z^3}\right) + 10\left(z - \frac{1}{z}\right)$$

but $z^n - \dfrac{1}{zn} = 2i\sin n\theta$

$$\therefore\ (2i\sin\theta)^5 = 2i\{\sin 5\theta - 5\sin 3\theta + 10\sin\theta\}$$

$$\sin^5\theta \qquad = \frac{1}{16}\{\sin 5\theta - 5\sin 3\theta + 10\sin\theta\}$$

2 Express $\cos 5\theta$ in terms of powers of $\cos\theta$.
By De Moivre's theorem

$$\cos 5\theta + i\sin 5\theta = (\cos\theta + i\sin\theta)^5$$

$$= (c + i\,s)^5$$

where $c = \cos\theta$, $\quad s = \sin\theta$
expanding the right hand side.

$\cos 5\theta + i\sin 5\theta = c^5 + 5c^4\,i\,s + 10c^3(i\,s)^2 + 10c^2(i\,s)^3$

$$+ 5\,c(i\,s)^4 + (i\,s)^5$$
$$= c^5 + 5i\,c^4\,s - 10\,c^3\,s^2 - 10i\,c^2\,s^3 + 5\,c\,s^4 + i\,s^5$$

equating real terms on both sides

$$\cos 5\theta = c^5 - 10\,c^3\,s^2 + 5\,c\,s^4$$
$$= \cos^5\theta - 10\cos^3\theta\sin^2 + 5\cos\theta\sin^4\theta$$
$$= \cos^5\theta - 10\cos^3\theta(1-\cos^2\theta) + 5\cos\theta(1-\cos^2\theta)^2$$
$$= \cos^5\theta - 10\cos^3\theta + 10\cos^5\theta + 5\cos\theta - 10\cos^3\theta + 5\cos^5\theta$$
$$= \mathbf{16\cos^5\theta - 20\cos^3\theta + 5\cos\theta}$$

Practice questions 8

1 Express $\cos^6\theta$ in terms of cosines of multiples of θ.
2 Express $\sin^7\theta$ in terms of sines of multiples of θ.
3 Show that $\sin 4\theta = \sin\theta\{8\cos^3\theta - 4\cos\theta\}$.
4 Express $\sin 7\theta$ in terms of powers of $\sin\theta$.

4 TRIGONOMETRY

The general angle and the trigonometrical ratios

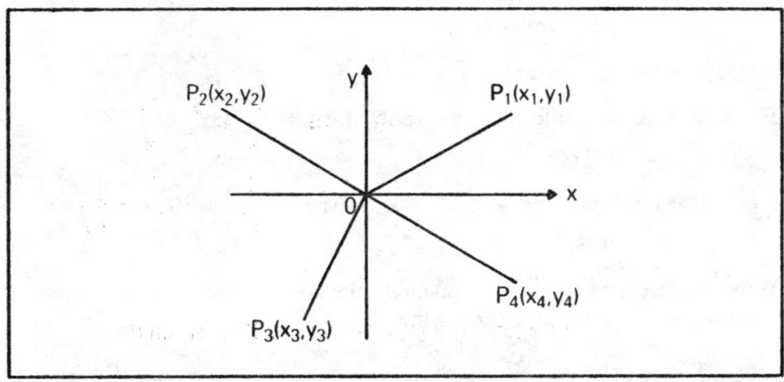

Angles measured from a given direction, Ox say, in an anticlockwise sense are regarded as positive and angles measured in the clockwise sense from Ox are regarded as negative.
The trigonometrical ratios of an angle $xOP = \theta°$, where P is any point with co-ordinates (x, y), are defined as follows.

$$\sin\theta = \frac{y}{OP}, \qquad \cos\theta = \frac{x}{OP}, \qquad \tan\theta = \frac{y}{x}$$

$$\operatorname{cosec}\theta = \frac{1}{\sin\theta}, \qquad \sec\theta = \frac{1}{\cos\theta}, \qquad \cot\theta = \frac{1}{\tan\theta}$$

Four possible positions of P are shown in the above figure, for P_1 both x_1 and y_1 are positive; for P_2, x_2 is negative and y_2 is positive, and so on. The ratios which are thus positive in each quadrant can be shown in the following diagram.

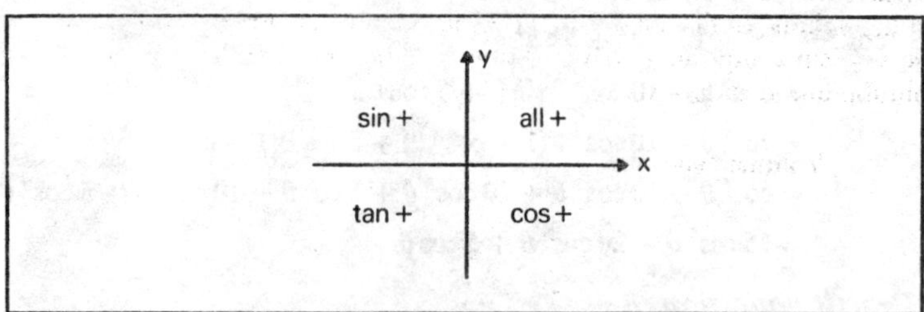

For any angle in the second quadrant, $\angle xOP_2 = \psi°$ say

$$\sin \psi° = \sin (180° - \psi°)$$
$$\cos \psi° = -\cos (180° - \psi°)$$
$$\tan \psi° = -\tan (180° - \psi°)$$

For any angle in the third quadrant, $\angle xOP_3 = \alpha°$ say

$$\sin \alpha° = \sin (\alpha° - 180°)$$
$$\cos \alpha° = \cos (\alpha° - 180°)$$
$$\tan \alpha° = \tan (\alpha° - 180°)$$

For any angle in the fourth quadrant, $xOP_4 = \phi°$ say

$$\sin \phi° = -\sin (360° - \phi°)$$
$$\cos \phi° = \cos (360° - \phi°)$$
$$\tan \phi° = -\tan (360° - \phi°)$$

When finding the trig. ratios of an angle always show the angle in relation to the co-ordinate axes in a sketch first.

Examples

1 Write down the values of the following (in surd form)

(i) $\sin 240°$ (ii) $\cos 330°$ (iii) $\tan(-150°)$

(iv) $\cot 135°$ (v) $\sec(-405°)$

(i)

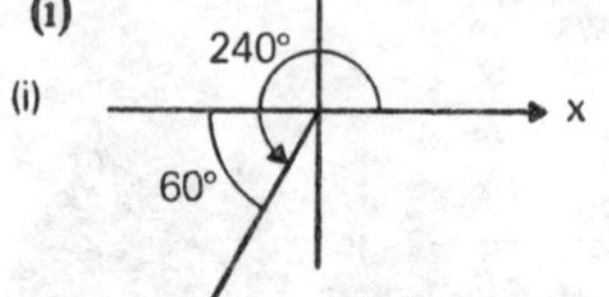

$$\sin 240° = -\sin (240° - 180°)$$
$$= -\sin 60°$$
$$= -\frac{\sqrt{3}}{2}$$

(ii)

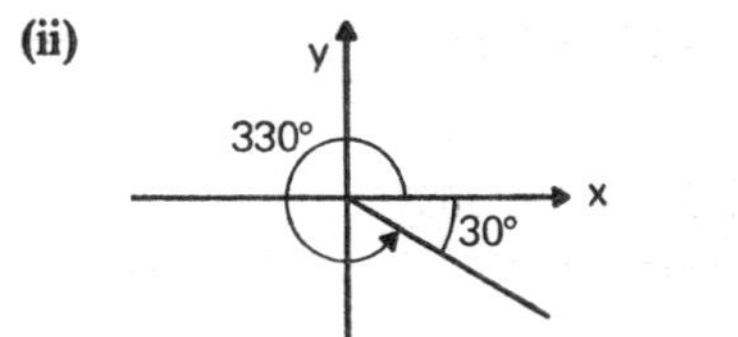

$$\cos 330° = \cos(360° - 330°)$$
$$= \cos 30°$$
$$= \frac{\sqrt{3}}{2}$$

(iii)

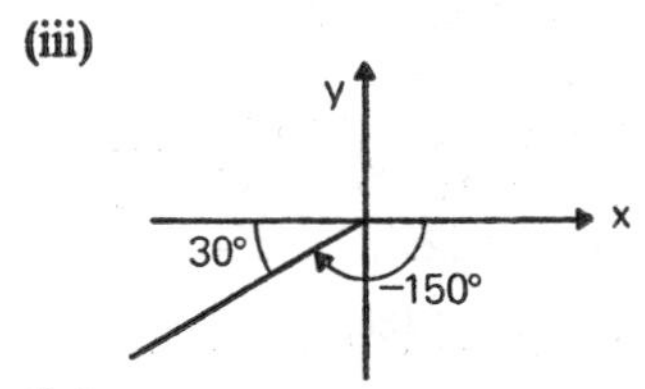

$$\tan(-150°) = \tan(180° - 150°)$$
$$= \tan 30°$$
$$= \frac{1}{\sqrt{3}}$$

(iv)

$$\cot 135° = -\cot(180° - 135°)$$
$$= -\cot 45°$$
$$= -1$$

(v)

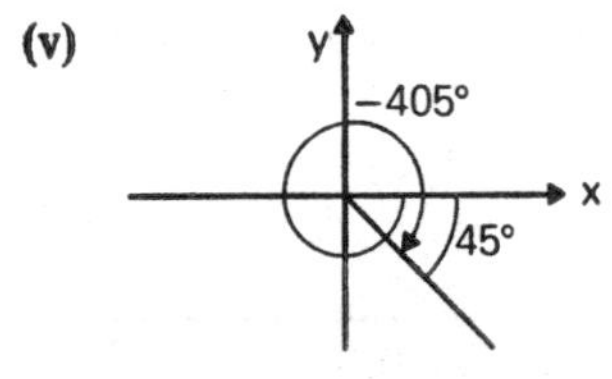

$$\sec(-405°) = \sec[-(405° - 360°)]$$
$$= \sec(-45°)$$
$$= \sec 45°$$
$$= \sqrt{2}$$

2 Find the value of θ from 0° to 360° which satisfy the following equations.

(i) $\cos\theta = -\dfrac{\sqrt{3}}{2}$ (ii) $\sin(\theta - 30°) = \dfrac{1}{\sqrt{2}}$

(i) the positive acute angle α such that $\cos\alpha = \sqrt{3}/2$ is 30° since $\cos\theta$ is negative θ lies in either the second or the third quadrant

$$\left.\begin{aligned} \therefore\ \theta &= 180° - 30° = 150° \\ \text{or}\ \theta &= 180° + 30° = 210° \end{aligned}\right\}$$

(ii) the positive acute angle α such that $\sin\alpha = 1/\sqrt{2}$ is 45°, since $\sin(\theta - 30°)$ is positive $(\theta - 30°)$ lies in either the first or the second quadrant

$$\therefore\ (\theta - 30°) = 45°$$
$$\theta = 75°$$
$$\text{or}\ \theta - 30° = 180° - 45°$$

$$\theta - 30° = 135°$$

$$\theta = 165°$$

i.e. $\theta = 75°$ or $165°$

Practice questions 1

1 Write down the values of the following

(1) $\sin 135°$, (ii) $\tan(-60°)$, (iii) $\sec 150°$, (iv) $\cos(-210°)$

(v) $\sin 420°$, (vi) $\operatorname{cosec}(-120°)$, (vii) $\cot 225°$,

(viii) $\sin(-180°)$, (ix) $\cos(-360°)$, (x) $\tan 180°$

2 Solve the following equations for values of θ from 0° to 360°

(i) $\tan\theta = -1$ (ii) $\cos(\theta - 20°) = -\frac{1}{2}$

3 Solve the following equations for θ, giving all values lying in the interval $-180° \leq \theta \leq 180°$

(i) $2\cos^2\theta + \cos\theta = 0$ (ii) $2\sin^2\theta + \sin\theta = 1$

Circular measure – radians

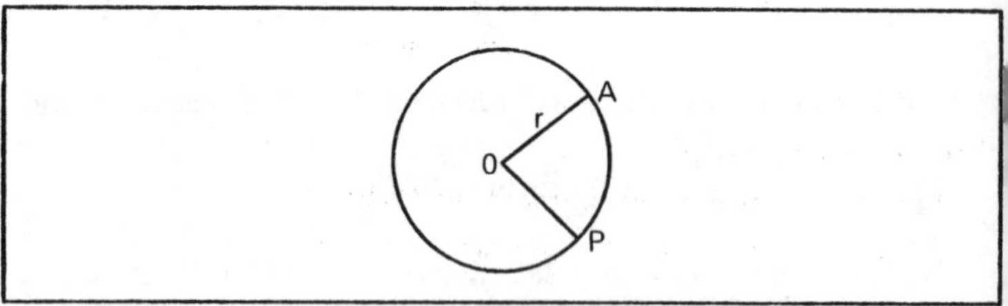

AP is an arc of a circle centre *O* and radius *r*. The size of the angle *AOP* may be defined by the ratio of the arc length *AP* to the radius *r*. In the particular case when the length of arc is equal to the radius of the circle the angle *AOP* is called a radian.

Thus in a circle of radius *r* an arc length of *kr* subtends an angle of *k* radians at the centre of the circle. Since the circumference of a circle is $2\pi r$ an angle of one revolution is equal to 2π radians.

i.e. $360° = 2\pi$ radians

The ratio arc length/radius is independent of the size of the circle and is a pure number. The concept of an angle has thus been generalised and the angle AOP may be defined as arc length/radius.
When θ is measured in radians the length of arc AP which subtends an angle θ at the centre is $r\theta$ and the area of the sector AOB is $\frac{1}{2}r^2\theta$.

Example
A chord BQ of a circle radius r subtends an angle α radians at any point on the remaining part of the circumference. Find the area of the minor segment cut off by the chord BQ.

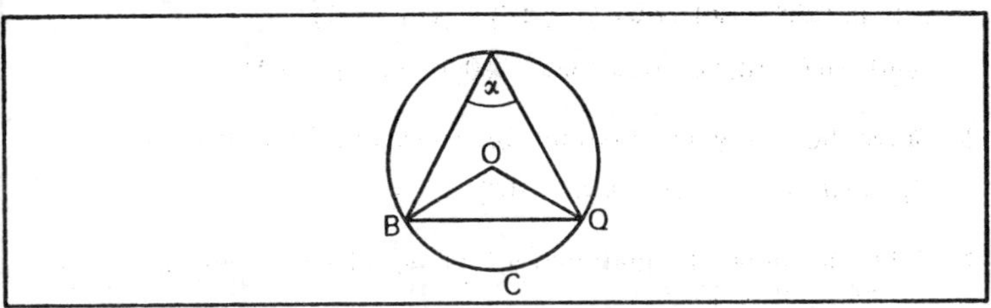

Let O be the centre of the circle, then $\angle BOQ = 2\alpha$

$$\text{area of the sector } OBCQ = \tfrac{1}{2}r^2 . 2\alpha = \alpha r^2$$

$$\text{area of triangle } OBQ = \tfrac{1}{2}OB . OQ \sin \angle BOQ$$

$$= \tfrac{1}{2}r^2 \sin 2\alpha$$

$$\text{area of minor segment } BCQ = \alpha r^2 - \tfrac{1}{2}r^2 \sin 2\alpha$$

$$= \tfrac{1}{2}\mathbf{r}^3 \,(\mathbf{2\alpha - \sin 2\alpha})$$

Practice questions 2

1 Convert to radians, leaving π in your answer:

(i) 90° (ii) 30° (iii) 45° (iv) 60° (v) 120°

(vi) 270° (vii) 300°

2 Convert to degrees the following angles measured in radians

(i) π (ii) $\frac{\pi}{4}$ (iii) $\frac{3\pi}{2}$ (iv) $\frac{4\pi}{3}$ (v) $\frac{\pi}{6}$

3 AD is a diameter of a circle radius r. P is a point on the circumference such that $\angle PAD = \theta$. Find:

(i) the perimeter,

(ii) the area of the region bounded by AP, AD and the minor arc PD.

Graphs of the trigonometric functions

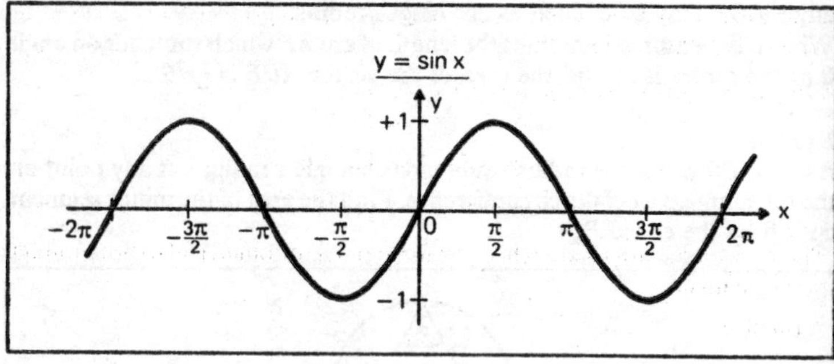

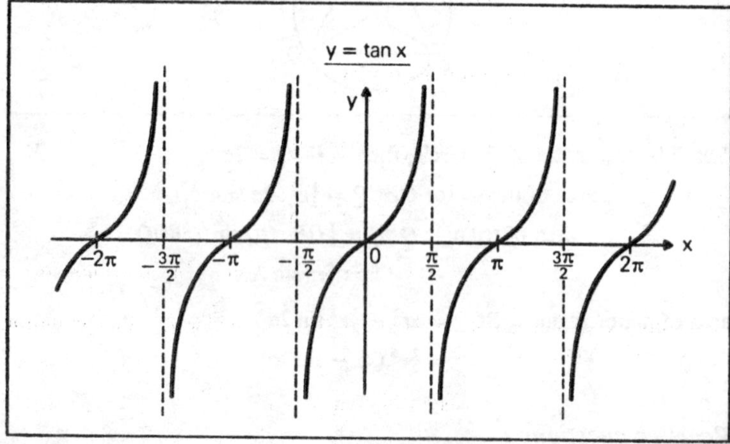

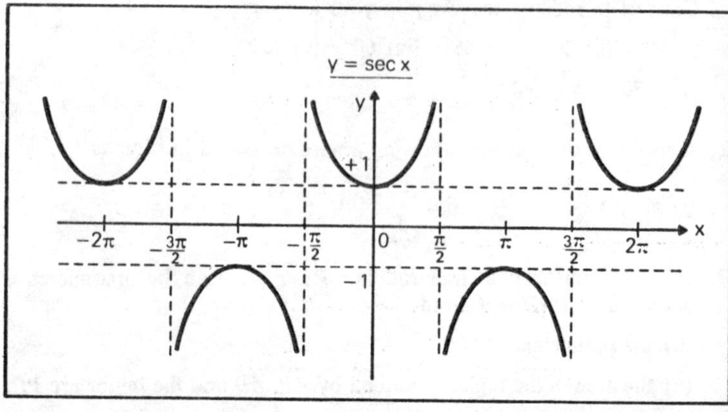

All the trigonometric functions are periodic.

Since $\cos x = \sin[(\pi/2) - x]$ and $\cot x = 1/\tan x$ the graphs of $\cos x$, $\cot x$ and $\operatorname{cosec} x$ can be deduced from the above three graphs. This is left as an exercise for the reader.

Pythagoras' theorem, trigonometrical forms

$$\sin^2\theta + \cos^2\theta = 1$$

$$\sec^2\theta = 1 + \tan^2\theta$$

$$\operatorname{cosec}^2\theta = 1 + \cot^2\theta$$

These formulae are useful when the form of trigonometrical expression has to be changed.

Examples

1

$$\int \sin^3 x\, dx$$

$$= \int \sin^2 x \sin x\, dx$$

$$= \int (1 - \cos^2 x)\sin x\, d x$$

$$= \int \sin x\, dx - \int \cos^2 x \sin x\, dx$$

$$= -\cos x + \tfrac{1}{3}\cos^3 x + k, \; k \text{ an arbitrary constant}$$

2 Solve, for values of x from 0 to 2π inclusive, the equation $\sec^2 x + 3\tan x + 1 = 0$.

$$\sec^2 x + 3\tan x + 1 = 0$$

$$1 + \tan^2 x + 3\tan x + 1 = 0$$

$$\tan^2 x + 3\tan x + 2 = 0$$

$$(\tan x + 1)(\tan x + 2) = 0$$

$$\text{either } \tan x = -1 \quad \text{or} \quad \tan x = -2$$

$$x = \frac{3\pi}{4}, \; \frac{7\pi}{4} \quad \text{or} \quad x = 2\pi \pm 1{\cdot}108$$

Important trigonometrical identities

$$\sin(A \pm B) = \sin A\cos B \pm \cos A\sin B$$

$$\cos(A \pm B) = \cos A\cos B \mp \sin A\sin B$$

$$\tan(A \pm B) = \frac{\tan A \pm \tan B}{1 \mp 1\text{b} \tan A \tan B}$$

By putting $A + B = P$ and $A - B = Q$ in the appropriate formulae above it may easily be deduced that

$$\sin P + \sin Q = 2\sin\frac{P+Q}{2}\cos\frac{P-Q}{2}$$

$$\sin P - \sin Q = 2\cos\frac{P+Q}{2}\sin\frac{P-Q}{2}$$

$$\cos P + \cos Q = 2\cos\frac{P+Q}{2}\cos\frac{P-Q}{2}$$

$$\cos P - \cos Q = -2\sin\frac{P+Q}{2}\cos\frac{P-Q}{2}$$

The double angle identities are

$$\sin 2x = 2\sin x\cos x$$

$$\cos 2x = \cos^2 x - \sin^2 x$$

$$= 2\cos^2 x - 1$$

$$= 1 - 2\sin^2 x$$

$$\tan 2x = \frac{2\tan x}{1-\tan^2 x}$$

$$\sin 2x = \frac{2\tan x}{1+\tan^2 x}$$

$$\cos 2x = \frac{1-\tan^2 x}{1+\tan^2 x}$$

If $t = \tan\frac{x}{2}$ then

$$\sin x = \frac{2t}{1+t^2}$$

$$\cos x = \frac{1-t^2}{1+t^2}$$

$$\tan x = \frac{2t}{1+t^2}$$

Use of these formulae have been shown in a variety of examples in this book.

Examples

1 If A, B and C are the interior angles of any triangle show that

$$\sin A + \sin B = 2\cos\frac{A-B}{2}\cos\frac{C}{2}$$

and hence express $\sin A + \sin B + \sin C$ as a product of the cosines of half angles of the triangle.

$$A + B + C = 180° \quad \text{(angles of a triangle)}$$

$$\sin A + \sin B = 2\sin\frac{A+B}{2}\cos\frac{A-B}{2}$$

$$= 2\sin\frac{180° - C}{2}\cos\frac{A-B}{2}$$

$$= 2\cos\frac{C}{2}\cos\frac{A-B}{2}$$

$$\sin A + \sin B + \sin C = 2\cos\frac{C}{2}\cos\frac{A-B}{2} + \sin C$$

$$= 2\cos\frac{C}{2}\cos\frac{A-B}{2} + 2\sin\frac{C}{2}\cos\frac{C}{2}$$

$$= 2\cos\frac{C}{2}\left[\cos\frac{A-B}{2} + \sin\frac{C}{2}\right]$$

$$= 2\cos\frac{C}{2}\left[\cos\frac{A-B}{2} + \sin\frac{180° - A - B}{2}\right]$$

$$= 2\cos\frac{C}{2}\left[\cos\frac{A-B}{2} + \sin\frac{(90 - A + B)}{2}\right]$$

$$= 2\cos\frac{C}{2}\left[\cos\frac{A-B}{2} + \cos\frac{A+B}{2}\right]$$

$$= 2\cos\frac{C}{2}\,2\cos\frac{A}{2}\cos\left(-\frac{B}{2}\right)$$

$$= 4\cos\frac{A}{2}\cos\frac{B}{2}\cos\frac{C}{2}$$

2 Prove that $\sin^2(P+Q) - \sin^2 P - \sin^2 Q = 2\sin P\sin Q\cos(P+Q)$

L.H.S. $= \sin^2(P+Q) - \sin^2 P - \sin^2 Q$

$= (\sin P\cos Q + \cos P\sin Q)^2 - \sin^2 P - \sin^2 Q$

$= \sin^2 P\cos^2 Q + 2\sin P\cos Q\cos P\sin Q + \cos^2 P\sin^2 Q$
$\quad - \sin^2 P - \sin^2 Q$

$= 2\sin P\cos Q\cos P\sin Q - \sin^2 P + \sin^2 P\cos^2 Q - \sin^2 Q$
$\quad + \cos^2 P\sin^2 Q$

$= 2\sin P\cos Q\cos P\sin Q - \sin^2 P(1 - \cos^2 Q)$
$\quad - \sin^2 Q(1 - \cos^2 P)$

$= 2 \sin P \cos Q \cos P \sin Q - \sin^2 P \sin^2 Q - \sin^2 Q \sin^2 P$

$= 2 \sin P \cos Q \cos P \sin Q - 2 \sin^2 P \sin^2 Q$

$= 2 \sin P \sin Q(\cos P \cos Q - \sin P \sin Q)$

$= 2 \sin P \sin Q \cos(P + Q)$

$=$ R.H.S.

Practice questions 3

1 Prove the identities

(i) $\sin 4A \cos 2A + \cos 4A \sin 2A = 2 \sin 3A \cos 3A.$

(ii) $3 \sin 2\alpha \cos 2\alpha = 2 \cos 3\alpha \sin^3 \alpha + 2 \sin 3\alpha \cos^3 \alpha.$

2 If A, B and C are the angles of triangles ABC prove that
$\cos A \cos B \cos C = \sin^2 A + \cos^2 A \cos(B - C) - 1$

Trigonometric equations; general solutions

If α radians is one solution of the equation

$$\sin x = k, \quad (-1 \leq k \leq 1),$$

the general solution is

$$x = n\pi + (-1)^n\alpha, \quad (n \text{ any integer})$$

If β radians is one solution of the equation

$$\cos x = k, \quad (-1 \leq k \leq 1),$$

the general solution is

$$x = 2n\pi \pm \beta, \quad (n \text{ any integer})$$

If α radians is one solution of the equation

$$\tan x = k,$$

the general solution is

$$x = n\pi + \alpha, \quad (n \text{ any integer})$$

Example

1 Find in radians the general solution to the equation

$$2 \sin^2 \theta = \cos \theta + 1$$

$$2 \sin^2 \theta = \cos \theta + 1$$

$$2(1 - \cos^2 \theta) = \cos \theta + 1$$

$$2 - 2 \cos^2 \theta = \cos \theta + 1$$

$$2\cos^2\theta + \cos\theta - 1 = 0$$

$$(2\cos\theta - 1)(\cos\theta + 1) = 0$$

either $\cos\theta = \frac{1}{2}$ or $\cos\theta = -1$

one value of θ when $\cos\theta = \frac{1}{2}$ is $\frac{\pi}{3}$ and the general value is

$$0 = 2n\pi \pm \frac{\pi}{3} \quad (n \text{ any integer})$$

one value of θ when $\cos\theta = 1$ is π and the general value is

$$\theta = 2n\pi \pm \pi \quad (n \text{ any integer})$$

the general solution is in two parts,

$$\boldsymbol{\theta = 2n\pi \pm \frac{\pi}{3}, \quad \theta = 2n\pi \pm \pi \quad (\text{n any integer})}$$

Practice questions 4

1 Find, in radians, the general solution of the equations

(i) $(2\sin x + 1)(\tan x - 1) = 0$

(ii) $\sin 2\theta = \sqrt{3}\cos\theta$

2 Find the general solution, correct to the nearest minute, for the equation

$$4\sec^2 x - 7 = 4\tan x$$

Equations of the form a *cos* x + b *sin* x = c

Method 1

$a\cos x + b\sin x = c$, where a, b and c are known constants.
divide both sides of the equation by $+\sqrt{a^2 + b^2}$

$$\frac{a}{\sqrt{a^2+b^2}}\cos x + \frac{b}{\sqrt{a^2+b^2}}\sin x = \frac{c}{\sqrt{a^2+b^2}} \quad (1)$$

Let $\frac{a}{\sqrt{a^2+b^2}} = \cos\alpha$ and $\frac{b}{\sqrt{a^2+b^2}} = \sin\alpha$

a value of the angle α can then be determined
(1) becomes

$$\cos x\cos\alpha + \sin x\sin\alpha = \frac{c}{\sqrt{a^2+b^2}}$$

$$\cos(x - \alpha) = \frac{c}{\sqrt{a^2+b^2}}$$

from whence the general solution may be obtained.

Method 2

$$a\cos x + b\sin x = c$$

Let $t = \tan\frac{1}{2}x$ then

$\cos x = \dfrac{1-t^2}{1+t^2}$ and $\sin x = \dfrac{2t}{1+t^2}$ and the equation becomes

$$a\frac{1-t^2}{1+t^2} + b\frac{2t}{1+t^2} = c$$

$$a(1-t^2) + 2bt = c(1+t^2)$$

this is a quadratic equation which can be solved for t and from whence the general solution may be obtained.

Example

Find the solutions lying between 0° and 360°, and the general solution of the equation

$$3\cos\theta - 4\sin\theta = 1$$

Method 1

$$3\cos\theta - 4\sin\theta = 1$$

$$\sqrt{3^2 + (-4)^2} = 5$$

dividing both sides by 5

$$\frac{3}{5}\cos\theta - \frac{4}{5}\sin\theta = \frac{1}{5} \qquad \text{(i)}$$

Let α be the angle such that

$$\cos\alpha = \frac{3}{5} \quad \text{and} \quad \sin\alpha = \frac{4}{5}$$

from table $\alpha = 53^\circ\,8'$

equation (i) may be re-written in the form

$$\cos\theta\cos\alpha - \sin\theta\sin\alpha = \frac{1}{5}$$

$$\cos(\theta + \alpha) = \frac{1}{5}$$

$$\cos(\theta + 53^\circ\,8') = \frac{1}{5}$$

$$\theta + 53° \, 8' = \cos^{-1} 0{\cdot}2$$

from tables $\cos^{-1} 0{\cdot}2 = 78° \, 28'$

For $0 \le \theta \le 360°$, $\theta + 53° \, 8' = 78° \, 28'$ or $(360° - 78° \, 28')$

$$\text{either } \theta = 78° \, 28' - 53° \, 8'$$

$$= 25° \, 20'$$

$$\text{or } \theta = 360° - 78° \, 28' - 53° \, 8'$$

$$= 18° \, 24'$$

for the general solution,

$$\theta + 53° \, 8' = 2n \times 180° \pm 78° \, 28', \quad n \text{ an integer}$$

$$\theta = n \times 360° - 53° \, 8' \pm 78° \, 28', \quad n \text{ an integer}$$

the general solution may be written as

$$\theta = n \times 360° - 53° \, 8' + 78° \, 28' \quad \text{and} \quad \theta = n \times 360° - 53° \, 8' - 78° \, 28'$$

$$\theta = n \times 360° - 25° \, 20' \quad \text{and} \quad \theta = n \times 360° - 131° \, 36'$$

Method 2

$$3 \cos \theta - 4 \sin \theta = 1$$

Let $\tan \frac{\theta}{2} = t$

$$\text{then } \sin \theta = \frac{2t}{1 + t^2} \quad \text{and} \quad \cos \theta = \frac{1 - t^2}{1 + t^2}$$

and the general equation may be rewritten in the form

$$3 . \frac{1 - t^2}{1 + t^2} - 4 . \frac{2t}{1 + t^2} = 1$$

$$3 - 3t^2 - 8t = 1 + t^2$$

$$4t^2 + 8t - 2 = 0$$

$$2t^2 + 4t - 1 = 0$$

$$t = \frac{-4 \pm \sqrt{16 + 8}}{4}$$

$$= -1 \pm \frac{\sqrt{6}}{2}$$

$$= -1 \pm \frac{2{\cdot}449}{2}$$

$$= -1 \pm 1{\cdot}2245$$

$$\text{either } t = -1 + 1{\cdot}2245$$
$$t = 0{\cdot}2245$$
$$\tan\frac{\theta}{2} = 0{\cdot}2245$$
$$\frac{\theta}{2} = 12^\circ\,40' \quad \text{(from tables)}$$

giving the general solution $\frac{\theta}{2} = n \times 180^\circ + 12^\circ\,40'$

$$\theta = n \times 360^\circ + 25^\circ\,20'$$
$$\text{or } t = -1 - 1{\cdot}2245$$
$$= -2{\cdot}2245$$
$$\tan\frac{\theta}{2} = -2{\cdot}2245$$
$$\frac{\theta}{2} = -65^\circ\,48' \quad \text{(using tables)}$$

giving the general solution $\frac{\theta}{2} = n \times 180^\circ - 65^\circ\,48'$

$$\theta = n \times 360^\circ - 131^\circ\,36'$$

the general solution is thus

$\theta = n \times 360^\circ + 25^\circ\,20'$ and $\theta = 360^\circ - 131^\circ\,36'$ (as in the first method)

Practice questions 5

1 Find (a) the values of θ in the range $0 < 0 < 360^\circ$, (b) the general solution when θ satisfies the equation

$$3\cos\theta + 4\sin\theta = 1$$

2 Find the general solution of the equation

$$5\sin\theta - 12\cos\theta = 6{\cdot}5$$

Small angles

For $0 < \theta < \frac{\pi}{2}$, $\sin\theta < \theta < \tan\theta$

When θ is small, $\sin\theta \simeq \theta \simeq \tan\theta$

and $\lim_{\theta \to 0} \frac{\sin \theta}{\theta} = 1$

When θ is small, $\cos \theta \simeq 1 - \frac{1}{2}\theta^2$

and $$\sin \theta \simeq \theta - \frac{\theta^3}{3!}$$

Example
Without using tables or calculators find an approximate value of the acute angle θ which satisfies the equation

$$\sin \theta = 0{\cdot}5002$$

Now $$\sin \alpha = 0{\cdot}5$$

$$\alpha = \frac{\pi}{6}$$

Let $\theta = \frac{\pi}{6} + \varepsilon$ where ε is small

$$\sin\left(\frac{\pi}{6} + \varepsilon\right) = 0{\cdot}5002$$

$$\sin\frac{\pi}{6}\cos\varepsilon + \cos\frac{\pi}{6}\sin\varepsilon = 0{\cdot}5002$$

$$\frac{1}{2}\cos\varepsilon + \frac{\sqrt{3}}{2}\sin\varepsilon = 0{\cdot}5002 \qquad (1)$$

when ε is small $\cos\varepsilon \simeq 1$ and $\sin\varepsilon \simeq \varepsilon$

here we may write as an approximation to (1)

$$\frac{1}{2} + \frac{\sqrt{3}}{2}\varepsilon = 0{\cdot}5002$$

$$\varepsilon = \frac{2}{\sqrt{3}} \times 0{\cdot}0002$$

$$= 0{\cdot}00023$$

$$\therefore\ \theta = \frac{\pi}{6} + 0{\cdot}00023$$

Practice questions 6

1 Without using tables or calculators find an approximate value of the

acute angle θ which satisfies the equation

(a) $\sin\theta = 0{\cdot}499$

(b) $\cos\theta = 0{\cdot}5001$

Check your answers by using a calculator.

5 CO-ORDINATE GEOMETRY

The straight line

If A is the point (x_1, y_1) and B is the point (x_2, y_2)

then $AB^2 = (x_2 - x_1)^2 + (y_2 - y_1)^2$;

the gradient of AB is $\dfrac{y_2 - y_1}{x_2 - x_1}$;

the equation of the straight line which passes through A and B is

$y - y_1 = m(x - x_1)$ where m is the gradient of AB;

the co-ordinates of the point dividing AB in the ratio $\lambda : w$ are

$$\left(\frac{\lambda x_2 + wx_1}{\lambda + w}, \ \frac{\lambda y_2 + wy_1}{\lambda + w}\right),$$

and when $\lambda = w = 1$, the co-ordinates of the mid-point of AB are

$$\left(\frac{x_1 + x_2}{2}, \ \frac{y_1 + y_2}{2}\right)$$

The angle θ between two straight lines with gradients m_1 and m_2 is given by

$$\tan\theta = \frac{m_1 - m_2}{1 + m_1 m_2}$$

When the lines are perpendicular $m_1 m_2 = -1$.
The length of the perpendicular from the point (X, Y) to the straight line $ax + by + c = 0$ is

$$\pm\frac{aX + bY + c}{\sqrt{a^2 + b^2}}$$

When the numerical value only of the length is required the $\pm$ signs may be ignored.But when comparing the lengths of perpendiculars from two or more points to the line, all points on the same side of the line will have the same sign. This is illustrated by the following example. Are the points (2, -1) and (3, 5) on the same side of the line

$$x - 3y + 6 = 0?$$

perpendicular from (2, −1) to $x - 3y + 6 = 0$ is

$$\frac{2 - 3(-1) + 6}{\sqrt{1^2 + (-3)^2}}$$

$$= \frac{11}{\sqrt{10}}$$

perpendicular from (3, 5) to $x - 3y + 6 = 0$ is

$$\frac{3 - 3(5) + 6}{\sqrt{10}}$$

$$= \frac{9 - 15}{\sqrt{10}}$$

$$= \frac{-6}{\sqrt{10}}$$

Hence the points are on opposite sides of the line.

The equations of the bisectors of the angles between the lines

$a_1x + b_1y + c_1 = 0$ and $a_2x + b_2y + c_2 = 0$ are

$$\frac{a_1x + b_1y + c_1}{\sqrt{a_1^2 + b_1^2}} = \pm\frac{a_2x + b_2y + c_2}{\sqrt{a_2^2 + b_2^2}}$$

Equation of a line passing through the point of intersection of the lines

$a_1x + b_1y + c_1 = 0$ and $a_2x + b_2y + c_2 = 0$

is $a_1x + b_1y + c_1 + k(a_2x + b_2y + c_2) = 0$, k a constant.

Examples

1 A and B are the points (3, 4) and (5, 9) respectively. Find the co-ordinates of the point which

(i) divides AB internally in the ratio 2 : 7

(ii) divides AB externally in the ratio 4 : 3.

(i) Let $P(X, Y)$ be the required point, then

$$X = \frac{2 \times 5 + 7 \times 3}{2 + 7} = \frac{31}{9}$$

and

$$Y = \frac{2 \times 9 + 7 \times 4}{2 + 7} = \frac{46}{9}$$

$P\left(\frac{31}{9}, \frac{46}{9}\right)$ divides AB internally in the ratio 2 : 7.

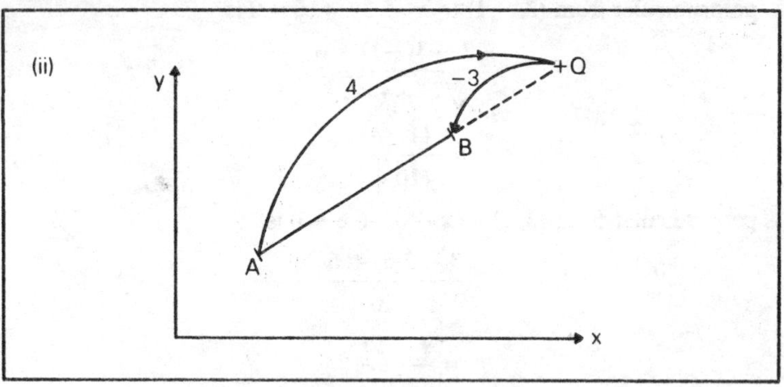

(ii) Let $Q(\alpha, \beta)$ divide AB externally in the ratio 4 : 3. In this case the λ and w of the general formula take on the values

$$\lambda = 4 \quad w = -3 \quad \text{and}$$

$$\alpha = \frac{4 \times 5 - 3 \times 3}{4 - 3} = 11$$

and

$$\beta = \frac{4 \times 9 - 3 \times 4}{4 - 3} = 24$$

Q(11, 24) divides AB externally in the ratio 4:3.

2 P is the point of intersection of the line $y = mx$ and the line passing through A (4, −4) which makes an angle of $\pi^c/4$ with the first line. Find the locus of P as m varies.

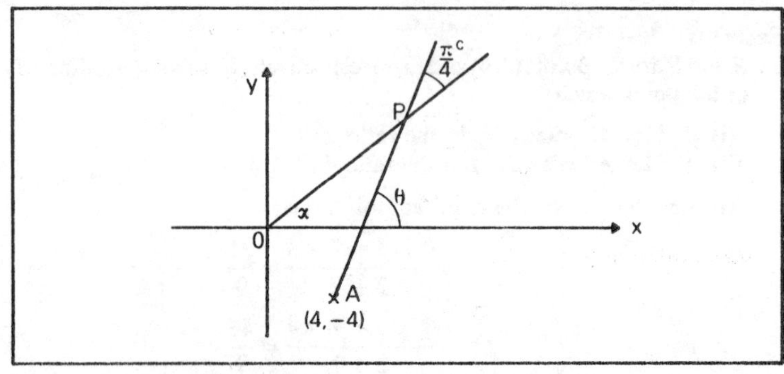

Let co-ordinates of P be X and Y when $\tan \alpha = m$
Let gradient of AP be $\tan \theta$

Then $\theta = \alpha + \dfrac{\pi}{4}$

$$\tan\theta = \tan\left(\alpha + \frac{\pi}{4}\right)$$

$$= \frac{\tan\alpha + \tan\dfrac{\pi}{4}}{1 - \tan\alpha\tan\dfrac{\pi}{4}}$$

$$= \frac{\tan\alpha + 1}{1 - \tan\alpha}$$

$$= \frac{m+1}{1-m}$$

equation of the line AP is

$$y + 4 = \frac{m+1}{1-m}(x - 4)$$

At the point of intersection, $P(X, Y)$

$$Y = mX \text{ (i)} \quad \text{and}$$

$$Y + 4 = \frac{m+1}{1-m}(X - 4) \text{ (ii) simultaneously}$$

Subtracting equation (i) from equation (ii)

$$4 = \frac{m+1}{1-m}(X - 4) - mX$$

$$4(1 - m) = (m + 1)(X - 4) - m(1 - m)X$$

$$4 - 4m = mX + X - 4m - 4 - mX + m^2X$$

$$8 = X(1 + m^2)$$

$$X = \frac{8}{1 + m^2}$$

Substituting in (i)

$$Y = \frac{8m}{1 + m^2} \quad \text{(ii)}$$

as m varies the parametric equations of the locus are

$$x = \frac{8}{1 + m^2}, \qquad x = \frac{8m}{1 + m^2}$$

$$\text{i.e. } x = \frac{8}{1+m^2} \qquad \text{and } y = mx$$

$$x + xm^2 = 8$$

$$m^2 = \frac{8-x}{x} \quad \text{and } y^2 = m^2x^2$$

eliminating m^2 we obtain

$$\frac{y^2}{x^2} = \frac{8-x}{x}$$

$$y^2 = 8x \times x^2$$

$$x^2 - 8x + 16 + y^2 = 16$$

$$(x-4)^2 + y^2 = 16$$

Locus is a circle centre (4, 0) and radius 4

Practice questions 1

1 Q, R, S and T are the points with co-ordinates (5, 2), (6, −2), (3, 1) and (2, −3) respectively. QT and RS intersect at N. Find the ratio in which N divides (i) RS (ii) TQ.

2 A variable straight line passing through the point (3, 5) meets the x-axis at P and the y-axis at Q. R divides PQ externally in the ratio (2 : 1). Find the equation of the locus of R.

3 The co-ordinates of P, Q and R are (−4, 5) (−2, 9) and (−1, 6) respectively. Find the value of the acute angle between PQ and PR and the equation of the bisectors of angle QPR.

The circle

The equation of the circle centre (x_1, y_1) and with radius r is,

$$(x - x_1)^2 + (y - y_1)^2 = r^2$$

This equation can be rearranged to give the form

$x^2 + y^2 + 2gx + 2fy + c = 0$, the general equation of a circle.

This is an equation of the second degree in which

(i) the coefficients of x^2 and y^2 are equal

and (ii) there is no term in xy.

The gradient of the tangent at any point (x_1, y_1) on the circle $x^2 + y^2 + 2gx + 2fy + c = 0$ can be obtained by differentiating this equation with respect to x, giving

$$2x + 2y\frac{dy}{dx} + 2g + 2f\frac{dy}{dx} = 0$$

$$\frac{dy}{dx} = -\frac{x+g}{y+f}$$

the gradient at (x_1, y_1) is $-\dfrac{x_1+g}{y_1+f}$

hence the equation of the tangent to the circle at (x_1, y_1) is

$$y - y_1 = -\frac{x_1+g}{y_1+f}(x - x_1)$$

simplifying we obtain

$$xx_1 + yy_1 + gx + fy = x_1^2 + y_1^2 + gx_1 + fy_1$$

and since (x_1, y_1) is a point on the circle $x_1^2 + y_1^2 + 2gx_1 + 2fy_1 + c = 0$

hence the **equation of the tangent to the circle at (x_1, y_1) is**

$\mathbf{xx_1 + yy_1 + g(x + x_1) + f(y + y_1) + c = 0}$

The length of the tangents from the point (x_1y_1) to the circle $x^2 + y^2 + 2gx + 2fy + c = 0$ is found as follows:

The equation $x^2 + y^2 + 2gx + 2fy + c = 0$ may be re-written as

$$(x+g)^2 + (y+f)^2 = g^2 + f^2 - c$$

the centre of this circle is the point $(-g, -f)$

and the radius is $\sqrt{g^2 + f^2 - c}$.

Since the tangent to a circle is perpendicular to the radius at the point of contact we get

$$[\text{length of a tangent from } (x_1, y_1)]^2 = [(x_1+g)^2 + y_1 + f)^2] - (g^2 + f^2 - c)$$
$$= \mathbf{x_1^2 + y_1^2 + 2gx_1 + 2fy_1 + c}$$

Example

Find the equation of the circle which passes through the point $(2, -1)$ and which touches the line $x + 5y + 4 = 0$ at the point $(1, -1)$.

Let the equation of the circle be

$$x^2 + y^2 + 2gx + 2fy + c = 0$$

The equation of the tangent at $(1, -1)$ is

$$x - y + g(x+1) + f(y-1) + c = 0$$

$$\text{i.e. } x(g+1) + y(f-1) + g - f + c = 0$$

But the equation of this tangent is given as

$$x + 5y + 4 = 0$$

hence
$$\frac{g+1}{1} = \frac{f-1}{5} = \frac{g-f+c}{4} \qquad \text{(i)}$$

Also since (2, −1) lies on the circle

$$4 + 1 + 4g - 2f + c = 0$$

$$4g - 2f = -c - 5 \qquad \text{(ii)}$$

From (i)
$$5g + 5 = f - 1$$

$$5g - f = -6 \qquad \text{(iii)}$$

and
$$4g + 4 = g - f + c$$

$$3g + f = c - 4 \qquad \text{(iv)}$$

eliminating c from (ii) and (iv)

$$4g - 2f = -3g - f - 4 - 5$$

$$7g - f = -9 \qquad \text{(v)}$$

$$5g - f = -6 \qquad \text{(iii)}$$

giving $g = -\frac{3}{2}$ $f = -\frac{3}{2}$

and substituting these values in (iv), $c = -2$

Hence the equation of the circle is $x^2 + y^2 - 3x - 3y - 2 = 0$

Practice questions 2

1 Find the equation of the circle which passes through the point (2, 3) and which touches the line $3x + 4y - 17 = 0$ at the point (3, 2).

2 Find the condition such that the circle with equation $x^2 + y^2 + 2g_1x + 2f_1y + c_1 = 0$ should cut orthogonally the circle with equation $x^2 + y^2 + 2g_2x + 2f_2y + c_2 = 0$.

3 Find the equation of the circle having the line joining the points (−3, 2) and 4, 5) as diameter.

6 ANSWERS

Chapter 1

Practice questions 1

1 (i) $(5x + 4y^2)(25x^2 - 20xy^2 + 16y^4)$
(ii) $3b^2(3a - 6b^2)(9a^2 + 18ab^2 + 36b^4)$

2 (i) $-(x+y+z)(x-y)(y-z)(z-x)$
(ii) $(a^2+b^2+c^2+bc+ca+ab)(b-c)(c-a)(a-b)$
3 (i) $(x-3)(x+2)(2x+5)$ (ii) $(2x+1)(3x-2)(4x-1)$
4 remainder 8. $x = -\frac{2}{3}, -\frac{1}{2}, 3$

Practice questions 2

1 (i) 3 (ii) $4x^3 - 12x^2 + 9x - 1 = 0$
2 1, 13
3 $x^3 - 2x^2 + x + 4 = 0$
4 $x^3 - 9x^2 + 3x - 39 = 0$, 6
5 $x^3 - 4x^2 + 9x - 7 = 0$

Practice questions 3

1 (i) $\dfrac{3}{2x+1} + \dfrac{x+1}{x^2+3x+2}$ (ii) $\dfrac{3}{x-1} + \dfrac{4}{x-2} - \dfrac{1}{3x-1}$

(iii) $\dfrac{2}{x-3} + \dfrac{3}{(x-3)^2} + \dfrac{4}{x+2}$

2 (i) $2x - 1 + \dfrac{2}{x-3} + \dfrac{3}{x+4}$

(ii) $1 + \dfrac{2}{x+2} + \dfrac{x}{x^2+1}$

Practice questions 4

1 $360.2.\dfrac{20!}{(10!)^2}$ 3 4804 4 1028

Practice questions 5

1 $\dfrac{n(n+1)(n^2+5n+2)}{4}$ 2 $\dfrac{n(4n^2+6n-1)}{3}$

3 $\dfrac{n}{4}(n+1)(n+2)(n+3)$ 4 $\dfrac{n}{5}(n+1)(n+2)(n+3)(n+4)$

5 $\dfrac{n}{2n+1}$ 6 $n.2^{n+2} - 3.2^{n+1} + 6$

Practice questions 6

1 $\dfrac{2^5.5.7}{3^3}$ 2 $\dfrac{8}{5}$, $-7{\cdot}2$

3 $1 + 2c_1x + 3c_2x^2 + 4c_3x^3$, $(n+1)c_nx^n$
4 (a) 1·006 (b) $1{\cdot}280 \times 10^{-12}$
5 624

Practice questions 7

1 (a) $\frac{1}{2}+\frac{5}{4}x+\frac{25}{8}x^2+\frac{125}{16}x^3$

(b) $1+\frac{9}{2}x+\frac{27}{8}x^2-\frac{27}{16}x^3$

2 $2+9x+31x^2+97x^3, \quad |x|<\frac{1}{3}$

Chapter 2

Practice questions 1

1 (i) $6x^5$ (ii) $24x^3$ (iii) $15(3x+2)^4$

Practice questions 2

1 (i) $-24x^{-4}$ (ii) $2x^{-5/7}+6x^{1/2}$
(iii) $30(2x^2+x+5)(4x+1)$ (iv) $4x-9x^2-5x^4$

2 (i) $3(x-5)^2(5x^4+5x^2-8x)+2(x-5)(3x^5+5x^3-12x^2-20)$
(ii) $-\frac{2}{3}(x^2+3x-2)^{-5/3}(2x+3)$ (iii) $\frac{4(13-5x)}{(4x-1)^3}$

Practice questions 3

1 (i) $3\cos 3x$ (ii) $-8x^3\sin(2x^4)$ (iii) $4\sec^2 x\tan^3 x$
(iv) $3\sec(3x-2)\tan(3x-2)$ (v) $-21x^2\operatorname{cosec} 7x^3\cot 7x^3$

2 (i) $5\cos 5x\cos 2x-2\sin 5x\sin 2x$
(ii) $-2\operatorname{cosec}^2 2x\sin^3 x+3\sin^2 x\cos x\cot 2x$
(iii) $\frac{7\cos 4x\cos 7x+4\sin 7x\sin 4x}{\cos^2 4x}$
(iv) $24x\sin^3(3x^2-2)\cos(3x^2-2)$

Practice questions 4

1 (i) $\frac{5}{\sqrt{1-25x^2}}$ (ii) $\frac{-3}{\sqrt{1-(3x-1)^2}}$ (iii) $\frac{24x^2}{1+64x^6}$

2 (i) $\frac{8}{\sqrt{1-(8x-1)^2}}$ (ii) $\frac{3}{\sqrt{4-9x^2}}$ (iii) $\frac{30x}{9+25x^4}$

Practice questions 5

1 (i) $7e^{7x}$ (ii) $10xe^{(5x^2+2)}$ (iii) $\frac{16x-3}{8x^2-3x+1}$

(iv) $\frac{7}{7x-3}+\frac{4}{4x-3}$

2 (i) $\dfrac{7}{7x-5}+\dfrac{1}{x-2}-\dfrac{4}{5-4x}$

(ii) $\dfrac{3}{3x-4}-\dfrac{1}{x-1}-\dfrac{1}{2x-3}$

(iii) $3\cos 3x e^{\sin 3x}$

Practice questions 6

1 (i) $9\cosh 9x$ (ii) $2x\sinh x^2$ (iii) $8\,\mathrm{sech}^2\, 8x$

2 (i) $\dfrac{8}{\sqrt{64x^2+1}}$ (ii) $\dfrac{10x}{\sqrt{25x^4-1}}$ (iii) $\dfrac{6x}{1-9x^4}$

Practice questions 7

1 (ii) $(x^2+3x-1)^3(x-4)^5\left[\dfrac{3(2x+3)}{x^2+3x-1}+\dfrac{5}{x-4}\right]$

(ii) $(x^2+5)^{1/2}(3x^4-x^2-1)^3\left[\dfrac{x}{x^2+5}+\dfrac{6x(6x^2-1)}{3x^4-x^2-1}\right]$

Practice questions 8

1 $a^n m(m-1)(m-2)(m-3)\ldots(m-n+1)(ax+b)^{m-n}$

2 (i) $a^n \sin\left(ax+b+\dfrac{\pi}{2}n\right)$ (ii) $a^n \cos\left(ax+\dfrac{\pi}{2}n\right)$

3 $y_5 = 4\sqrt{2}\,e^x \cos\left(x+5\dfrac{\pi}{4}\right)$, $y_n = (\sqrt{2})^n e^x \cos\left(x+\dfrac{\pi}{4}n\right)$

Practice questions 9

1 $-\tan\theta$, $\dfrac{1}{12}\sec^4\theta\,\mathrm{cosec}\,\theta$

2 $\dfrac{t^2-1}{2t}$ 3 $\tan\theta$

Practice questions 10

1 $\dfrac{-x(x+2y)}{x^2+y^2}$ 2 $\dfrac{\cos(x+y)}{1-\cos(x+y)}$

3 $\dfrac{x+y}{2y-x}$, $\dfrac{3(2y^2-2xy-x^2)}{(2y-x)^3}$

Practice questions 11

1 $y=6x-10$ 2 $(3, -6-3\ln 3)$

Practice questions 12

1 Min at $x = -3$ and $x = 1$, Max at $x = 0$
2 Min at $x = \frac{1}{2}$, Max $x = \frac{1}{2}$, infl. at $x = 0$, $x = \frac{3}{4}$
3 $(-\ln 2, \frac{1}{3}2^{2\ 3})$, maximum
4 Min at $x = 0$, $x = \pi$, $x = 2\pi$

Max at $x = \dfrac{\pi}{3}$

5 Min at $x = 2{\cdot}75$
infl at $x = -1$ and at $x = 1{\cdot}5$

Practice questions 13

1 (a) $\frac{1}{2}(3)^{3\ 2}$ (b) $4\sqrt{2}$
3 $3 \sin 2t$

Practice questions 14

1

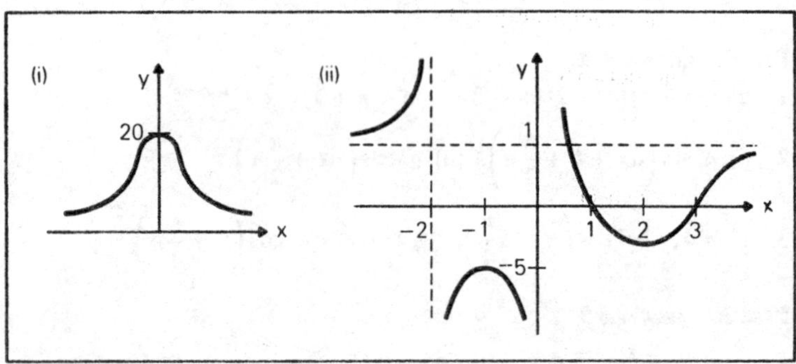

2 Turning points at $x = -\frac{2}{3}$ and at $x = 6$, point of inflexion at $x = 2\frac{2}{3}$

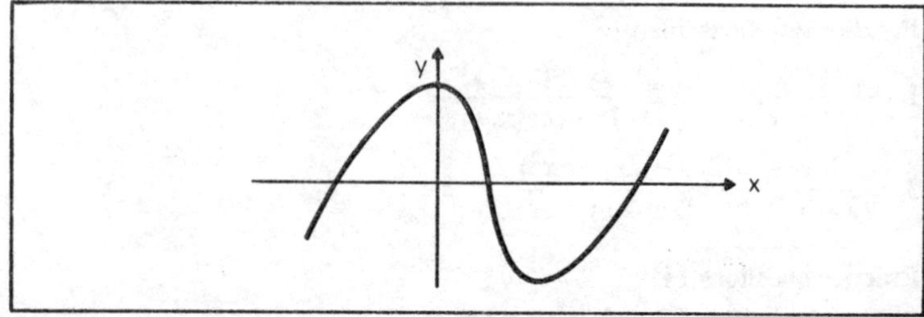

3

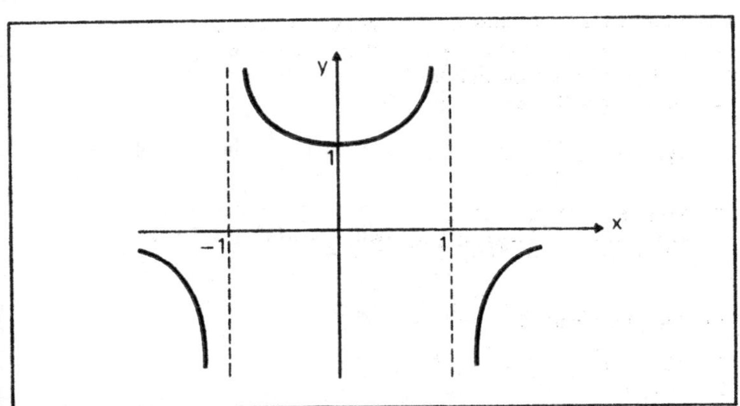

4 Maximum $(1, 1\frac{1}{2})$, Minimum $(-1, -1\frac{1}{2})$ points of inflexion $(0, 0)$, $(\sqrt{3}, \frac{3}{4}\sqrt{3})$ $(-\sqrt{3}, -\frac{3}{4}\sqrt{3})$
horizontal asymptote $y = 0$

Practice questions 15

1 limaçon

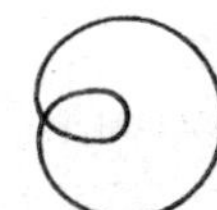

2 lemniscate

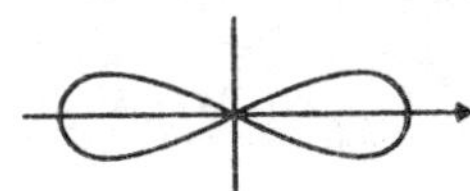

3 3 leaved rose

4 cardioid

Practice questions 16

k is an arbitrary constant.

1 (i) $\frac{1}{4}\tan^{-1}\frac{x}{4} + k$ (ii) $\frac{1}{4}\sin 4x + k$ (iii) $\frac{1}{7}e^{7x} + k$

(iv) $\frac{1}{5}\ln k \sec 5x$ (v) $\ln k(x^3 + 4)$ (vi) $\frac{1}{\sqrt{2}}\sin^{-1}\frac{x}{5} + 4$

(vii) $\frac{1}{\sqrt{42}}\tan^{-1}\left(x\sqrt{\frac{7}{6}}\right) + k$ (viii) $\frac{1}{4}\ln k(x^4 + 2x^2 - 5)$

(ix) $-\frac{1}{5}\cos 5x + k$ (xi) $\frac{5}{27}(3x + 2)^{9\ 5}$

2 (i) 0 (ii) $\frac{1}{3}(e^3 - 1)$ (iii) $\frac{64}{5}$

3 (i) $\frac{1}{2}\sinh 2x + k$ (ii) $\frac{1}{5}\sin^{-1}\left(x\sqrt{\frac{5}{6}}\right) + k$

(iii) $\cosh^{-1}\frac{x}{\sqrt{2}} + k$ (iv) $\frac{1}{2}\ln \cosh 2x + k$

Practice questions 17

k is an arbitrary constant.

1 $\frac{1}{2}x - \frac{1}{12}\sin 6x + k$ 2 $\frac{1}{4}\sin 2x + \frac{1}{2}x + k$

3 $-\frac{1}{14}\cos 7x - \frac{1}{6}\cos 3x + k$

4 $\frac{1}{2}\sin 2x + \frac{1}{10}\sin 5x + k$

5 $\frac{1}{3}\sin^3 x - \frac{2}{5}\sin^5 x + \frac{1}{7}\sin^7 x + k$

Practice questions 18

1 $\ln k\frac{x-2}{x}$ 2 $\frac{1}{2}\ln(x^2 + 1) + \tan^{-1} x + k$

3 $16\ln(x - 4) - 13\ln(x - 3) + 4$

Practice questions 19

1 $2\sqrt{x+1}\left[\frac{(x+1)^2}{5} - \frac{x+1}{3}\right] + k$

2 $-2(1-x)^{3\ 2}\left[\frac{1}{3} - \frac{2(1-x)}{5} + \frac{(1-x)^2}{7}\right] + k$

3 $\frac{9}{2}\left[\frac{1}{2} - \frac{\pi}{4}\right]$ 4 $\frac{625\pi}{16}$

5 $\mp\frac{\sqrt{9-x^2}}{9x} + k$

Practice questions 20

1 1 2 $\frac{\pi}{2}$ 3 1

Practice questions 21

1 $x \ln x - x + k$ 2 $x \sin x + \cos x + k$
3 $2x \sin x + 2\cos x - x^2 \cos x + c$

4 $\frac{1}{10}e^{-x}(3 \sin 3x - \cos 3x) + c$

5 $x \sin^{-1} x + \sqrt{1-x^2} + c$

6 $\frac{1}{2}(\sec x \tan x + \ln(\sec x + \tan x) + c$

7 $2 - \frac{5}{e}$

Practice questions 22

1 (i) $\frac{8}{15}$ (ii) $\frac{7}{8}\cdot\frac{5}{6}\cdot\frac{3}{4}\cdot\frac{1}{2}\cdot\frac{\pi}{2}$

2 $(n-1)I_n - \sec^{n-2} x \tan x + (n-2)I_{n-2}$

3 $I_n = \frac{1}{a}x^n e^{ax} - \frac{n}{a}I_{n-1}$

$\frac{3}{40}(3 + 3^{2\ 3})$

Practice questions 23

1 4 2 $3\frac{1}{12}$ 3 $\frac{\pi^2}{8} + \pi - 1$

Practice questions 24

1 3π 2 $\frac{1}{2}(3\pi - 8)$ 3 25 4 $5\pi - 8$

Practice questions 25

1 $\frac{4}{3}\pi r^3$ 2 $\pi(\ln 4 - 1)$

Practice questions 26

1 $e - \frac{1}{e}$ 2 $\ln 2 + \frac{3}{8}$ 3 $\frac{3}{2}$

4 24 5 $\frac{1}{2}\sinh 2a$

Practice questions 27

1 $\sqrt{2}\,a(e^\alpha - 1)$ 2 $\frac{3\pi}{2}$

Practice questions 28

1 $\frac{1}{3}\pi r^2$

2 $\frac{12\pi}{5}$

Practice questions 29

1 0

2 $\frac{1}{2}$

3 $\frac{8}{3}$

Chapter 3

Practice questions 1

1 (i) $3 + 17i$ (ii) $1 - 10i$ (iii) $-34 - 13i$ (iv) $\frac{13}{25} + \frac{34}{25}i$

1 (v) $\frac{191}{146} + \frac{47}{146}i$

2 $a = \frac{18}{25}$ $b = \frac{4}{25}$

3 $x = 22, \quad y = -3$

Practice questions 2

1 (a) $\sqrt{2}, \frac{7\pi}{4}$ (b) $8 \frac{\pi}{6}$ (c) $2, \frac{5\pi}{6}$

2 (a) $\sqrt{5} \arg 0{\cdot}464$ (b) $1 \arg \frac{3\pi}{2}$

Practice questions 3

1 $6 \operatorname{cis} 4\alpha$
2 $5 \operatorname{cis} (-4\theta)$
3 $\operatorname{cis} \alpha$
4 $\frac{2}{3} \operatorname{cis} 4\alpha$
5 $\operatorname{cis} \frac{\pi}{6}$
6 $\frac{1}{2} \operatorname{cis} \left(-\frac{\pi}{4}\right)$
7 $\sec \theta (\cos \theta + i \sin \theta)$

Practice questions 4

1 $3 - 3i$, $1 - 2i$, $5 + i$, $3 + 2i$

2 $\frac{1}{2}(-3 - 4\sqrt{3}) + \frac{i}{2}(-4 + 3\sqrt{3})$, $\frac{1}{2}(-3 + 4\sqrt{3}) + \frac{i}{2}(-4 - 3\sqrt{3})$

3 A is $5 \cos \alpha°$, B is $5 \operatorname{cis} (\alpha + 120°)$, C is $5 \operatorname{cis} (\alpha - 120°)$

Practice questions 5

1 (i) circle centre $(1, -5)$ radius 6
(ii) straight line, $y = x - 2$
(iii) circle, $x^2 + y^2 = 18$
2 Circle, $x^2 - x + y^2 + y - 6 = 0$
4 Circle

Practice questions 6

1 Straight line $3u + 1 = 0$
2 $|w| > 1$

3 circle centre $\left(\frac{5}{3}, 0\right)$ radius $\frac{4}{3}$

Practice questions 7

1 (a) $4 - 4i$ (b) $-128 + 128\sqrt{3}\,i$
3 $2, -1 + 3i\ -4, -1, -3i$

4 $\left(x^2 - 2x\cos\frac{\pi}{n} + 1\right)\left(x^2 - 2x\cos\frac{3\pi}{n} + 1\right)\ldots$

$\left(x^2 - 2x\cos\frac{(n-1)^{\pi}}{n} + 1\right)$

Practice questions 8

1 $\frac{1}{32}(\cos 6\theta + 6\cos 4\theta + 15\cos 2\theta + 10)$

2 $\frac{1}{64}(-\sin 7\theta + 7\sin 5\theta - 21\sin 3\theta + 35)$

4 $7\sin\theta - 56\sin^3\theta + 112\sin^5\theta - 64\sin^7\theta$

Chapter 4

Practice questions 1

1 (i) $\frac{1}{\sqrt{2}}$ (ii) $-\sqrt{3}$ (iii) $\frac{2}{\sqrt{3}}$ (iv) $-\frac{\sqrt{3}}{2}$ (v) $\frac{\sqrt{3}}{2}$

(vi) $-\frac{2}{\sqrt{3}}$ (vii) $\sqrt{2}$ (viii) 0 (ix) 1 (x) 0

2 (i) 135°, 315° (ii) 140°, 260°
3 (i) +120°, ±240°, ±90° (ii) −90°, 30°, 150°

Practice questions 2

1 (i) $\frac{\pi}{2}$ (ii) $\frac{\pi}{6}$ (iii) $\frac{\pi}{4}$ (iv) $\frac{\pi}{3}$ (v) $\frac{2\pi}{3}$ (vii) $\frac{5\pi}{3}$

2 (i) 180° (ii) 45° (iii) 270° (iv) 240° (v) 30°

3 (i) $2r(1 + \theta + \cos\theta)$ (ii) $\frac{1}{2}r^2(2r + \sin 2\theta)$

Practice questions 4

1 $x = n\pi + (-1)^n \dfrac{7\pi}{6}$ and $x = n\pi + \dfrac{\pi}{4}$

2 $x = n \times 180° - 26° 34'$ and $x = n \times 180° + 56° 19'$

Practice questions 5

1 $131° 36'$, $334° 40'$, $\theta = 53° 8' + n \times 360° \pm 78° 28'$
2 $\theta = 67° 23' + n \times 180° + (-1)^n 30°$

Chapter 5

Practice questions 1

1 17 : 7
2 $xy - 10x + 2y = 0$
3 $\dfrac{\pi}{4}$

Practice questions 2

1 $x^2 + y^2 - 12x - 12y + 47 = 0$
2 $2g_1g_2 + 2f_1f_2 = c_1 + c_2$
3 $x^2 - x + y^2 - 7y - 2 = 0$